An Introduction to Insect Pests and their Control

Peter D. Stiling

MACMILLAN
PUBLISHERS

First published 1985

Published by *Macmillan Publishers Ltd*
London and Basingstoke
Associated companies and representatives in Accra, Auckland, Delhi, Dublin, Gaborone, Hamburg, Harare, Hong Kong, Kuala Lumpur, Lagos, Manzini, Melbourne, Mexico City, Nairobi, New York, Singapore, Tokyo

ISBN 0 - 333 - 39240 - X

Printed in Hong Kong

Preface

Over the past 50 years a major portion of the world's population has been undernourished or starving. The World Bank has estimated that 65% of the people in developing countries receive at least 15% fewer calories than needed to function efficiently. This deficiency amounts to 10% of the caloric needs of that developing country and about 4% of the worlds caloric output. As 72% of the worlds population of 4000 million people are of developing nations, roughly half the world is hungry.

The productivity of a natural ecosystem is around 6 kg/ha so the world could support only 200 million people at most if we relied on its natural productivity. This figure was surpassed by the Middle Ages. We have little choice but to use energy-driven technology to raise the productivity of the land. Sometimes protein production is increased to as much as 50 times that of a natural system. The energy input for such technology is quite large, being at least 10% of the total national energy consumption of even the most developed nations. The 220 million people of the United States effectively 'eat' about 150 million tons of oil in this way: more than the entire caloric equivalent of the food provision for India's 600 million people. Energy is used in many ways including fertilizer and pesticide production, and irrigation.

Unfortunately, the OPEC decision in 1973 to increase oil prices by a factor of seven has closed the door to increases in production in many developing countries. Without production increases, preservation of existing yields becomes even more important, yet it is estimated that over a third of the potential world-wide crop yield is destroyed by pests, of which insect pests are by far the most predominant. Yet only a small fraction of the world's insects are pests; many more are beneficial in that they prey on or parasitise harmful insects or destroy important weeds. There is a clear need for management of our insect heritage; to control insect pests and to promote beneficial species.

Acknowledgements

The author and publishers wish to thank the following who have kindly given permission for the use of copyright material:

Table 4.1 McGraw-Hill Book Company for this table from *Destructive and Useful Insects* by R. L. Metcalf and W. P. Flint (3rd Edition) 1951.

Table 4.3 University Press of Mississippi for this table from an article by R. L. Metcalf in *Proceedings of the Summer Institute on Biological Control of Plant Insects and Disease*, 1974, edited by F. G. Maxwell and F. A. Harris.

Table 4.6 Entomological Society of America for this table from an article by E. F. Knipling in the *Journal of Economic Entomology*, issue no. 459, August 1955.

Table 5.1 University of California for this table from an article by Hall, *et al*, in the journal *California Agriculture*, issue 29, 1975.

Fig 1.3 Watson T. F., Moore, L. and Ware, G. W. (1975) *Practical Insect Pest Management*. Oxford: W. H. Freeman and Co. Copyright © 1975. All rights reserved

Fig 2.1 Data from Jones, D. P. (1973) *History of Entomology*, eds. Smith, Mittler and Smith. California, Palo Alto: Annual Reviews Inc.

Figs 3.1 and 3.4 *See figure captions*

Fig 4.2 Flint, M. L. and Van den Bosch, R. (1981) *Introduction to Integrated Pest Management*. New York: Plenum Press

Fig 4.8 After USA *Pesticide Registration Guidelines* (1975). Washington D.C: EPA.

Fig 4.9 After von Rumker *et al*. (1974) *Pesticide Study Series 2* Washington D.C: EPA.

Fig 4.10 Hassal, K. A. (1966) *Scientific Horticulture* **18**: 103–15

Fig 4.11 Van Emden, H. F. (1974) *Pest Control and its Ecology* Studies in Biology no. 50. London: Edward Arnold.

Fig 4.13 After Ratcliffe (1970) *Journal of Applied Ecology* **7**: 67–115.

Fig 4.16 Samways, M. J. (1981) *Biological Control of Pests and Weeds* Studies in Biology no. 132. London: Edward Arnold.

Fig 4.17 Varley, E. C., Gradwell, G. R. and Hassel, M. P. (1973) *Insect Population Ecology*. Oxford: Blackwell.

Cover photograph The Science Photo Library, London.

The publishers have made every effort to trace the copyright holders, but if they have inadvertently overlooked any, they will be pleased to make the necessary arrangements at the first opportunity.

Contents

For my parents

1 Introduction

1.1 The origins of pests

Derived from the Latin *pestis* for plague, 'pest' is a human invention used to describe plants (weeds), vertebrates, insects, mites, pathogens and other organisms which occur where we do not want them. Before man, there were no pests. Vast crop monocultures have aggravated the situation by promoting the build-up of huge pest populations. Transients and vagrant individuals reproduce quickly in a sea of succulent food, often in the absence of natural enemies. Insects, in particular, can exhibit a dramatic rate of increase if unchecked. One cabbage aphid (*Brevicoryne brassicae* L.), producing a new generation every two weeks, could annually produce 250 million tonnes of offspring, enough to circle the equator nose-to-tail one million times. One pair of house-flies (*Musca domestica*), in one year, could cover the earth to a depth of 15 m with their offspring (200 million million). Although very few of these progeny actually survive in nature, insects still occur in very high densities. One hectare of oats supports about 22 million fly larvae, whilst 222 million black bean aphids can be found on a hectare of sugar beet. The average density of insects is about 25 million per hectare of soil and 25 000 'in flight' above it. This compares with a human density on dry land of only 0.14 per hectare.

Most contemporary pest outbreaks occur when pest species are accidentally introduced to new habitats and new countries in the absence of their natural enemies. In this way the European corn borer, *Ostrinia nubitalis*, was introduced to the USA in 1909. More rarely a 'harmless' species reaches pest proportions after the introduction of new foodstuffs which are acceptable to that species. In 1824 the British entomologist Thomas Say named *Leptinotarsa decemlineata*, the Colorado potato beetle, as new to Science. Originally restricted to the local plant, buffalo-fur (of the potato family), in the Rocky Mountains, this species spread throughout the entire nation within 30 years owing to introduction of the potato. The insects spread at the rate of 140 km a year.

1.2 Pest damage

Even in the most highly-developed countries, losses to pests are staggering. In the United States during the 1960's, $7 billion was lost annually to pests, and a further $3 billion spent on management programmes. The total of $10 billion was equivalent to a quarter of the total crop value.

In less-developed countries the situation is much worse. In India, rats alone eat 10% of the grain, whilst in Africa $7 million worth of crops are lost to birds. In the latter case, much of the damage is due to the Quelea which is probably the single most important pest of the entire African continent. Although individuals in this species may eat only 2–3 g of food in a day, they often cause up to 8 times this amount of damage as they move through the fields. So far, frighteners, roost bombs and all other control techniques have failed against the Quelea.

In the Philippines, rats alone eat 90% of the rice, 80% of the corn and 50% of the sugar cane. Rodents are the most common of vertebrate pests although their apparent inability to vomit makes them susceptible to stomach poisons, the most common being the anti blood-coagulants warfarin, pival and fumarin. An added advantage here is that vitamin K, which increases blood-clotting ability, is readily available as an antidote for accidental cases of human poisoning. The

Table 1.1 Percentage losses of major crops to insects

Crops	Loss (%)
Rice	26.7
Wheat	5.0
Maize	12.4
Sorghum	9.6
Potatoes	6.5
Cassava	7.7
Sweet potatoes	8.9
Tomatoes	7.5
Soybeans	4.5
Peanuts	17.1
Palm oil	11.6
Copra	14.7
Cotton seed	11.0
Bananas	5.2
Citrus	8.3

disadvantage is that many rats have now become resistant to warfarin. Indeed, despite modern technology, pest species continue to be a problem. There seems to be no easy remedy. In 1932, US losses to insect pests were estimated to be 10%. In 1948 the estimate was again 10% and in 1969...10%?

1.3 The major types of pest

By far the biggest group of animals on earth is the class Insecta (Fig 1.1). There are more species of beetles than there are different vertebrates in the total chordate phylum. The weevil family alone is more speciose than the entire bird class. Since many of these insects are plant-feeders, or parasites of plant-feeders, insects feature prominently in pest management programmes (Fig 1.2).

Apart from the direct destruction of crops (up to 50% of the stored grain in many under-developed countries), insects cause indirect damage as vectors of plant and animal diseases.

Plant diseases can be categorised according to their effects:

1 **Necrosis** (death) includes rot, canker, spot, blight, yellowing and wilting;

2 **Hypoplasis** (under-development) includes dwarfing, rossetting, mosaic, chlorosis;

3 **Hyperplasis** (over-development) includes galls, blast, witches broom.

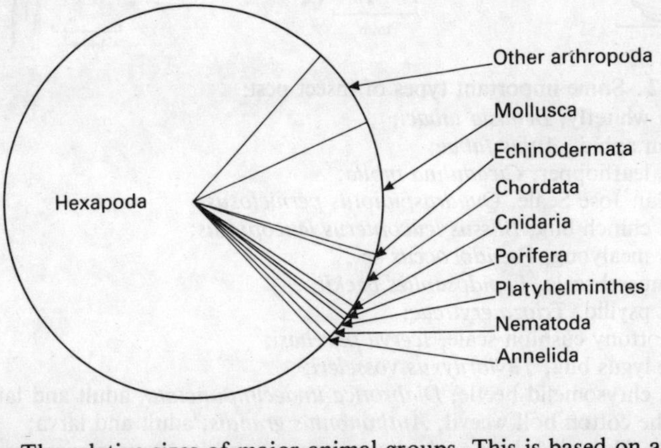

Fig 1.1 The relative sizes of major animal groups. This is based on an estimate of the number of living species in each group. The overwhelming dominance of the arthropods, in particular the hexapoda (consisting of the classes Insecta and Entognatha), is clear.

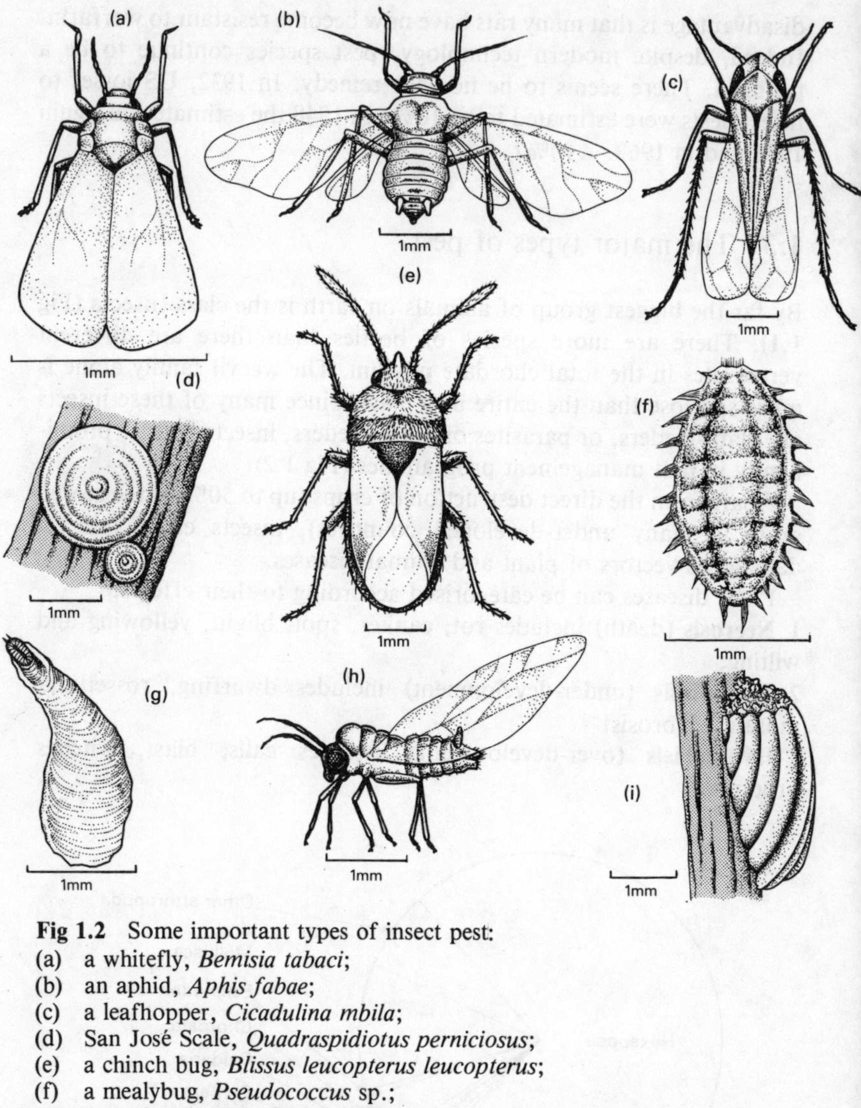

Fig 1.2 Some important types of insect pest:
(a) a whitefly, *Bemisia tabaci*;
(b) an aphid, *Aphis fabae*;
(c) a leafhopper, *Cicadulina mbila*;
(d) San José Scale, *Quadraspidiotus perniciosus*;
(e) a chinch bug, *Blissus leucopterus leucopterus*;
(f) a mealybug, *Pseudococcus* sp.;
(g) mussel scale, *Lepidosaphes beckii*;
(h) a psyllid, *Trioza erytreae*;
(i) cottony cushion scale, *Icerya purchasi*;
(j) a lygus bug, *Taylorilygus vosseleri*;
(k) a chrysomelid beetle, *Diabrotica undecimpunctata*, adult and larva;
(l) the cotton boll weevil, *Anthonomus grandis*, adult and larva;
(m) an adult click beetle, *Agriotes lineatus*, and its larva, a wireworm;
(n) the spring cankerworm, *Paleacrita vernata*, adult and larva;
(o) an armyworm, family Noctuidae, adult and larva;
(p) the hessian fly, *Mayetiola destructor*, adult and larva.

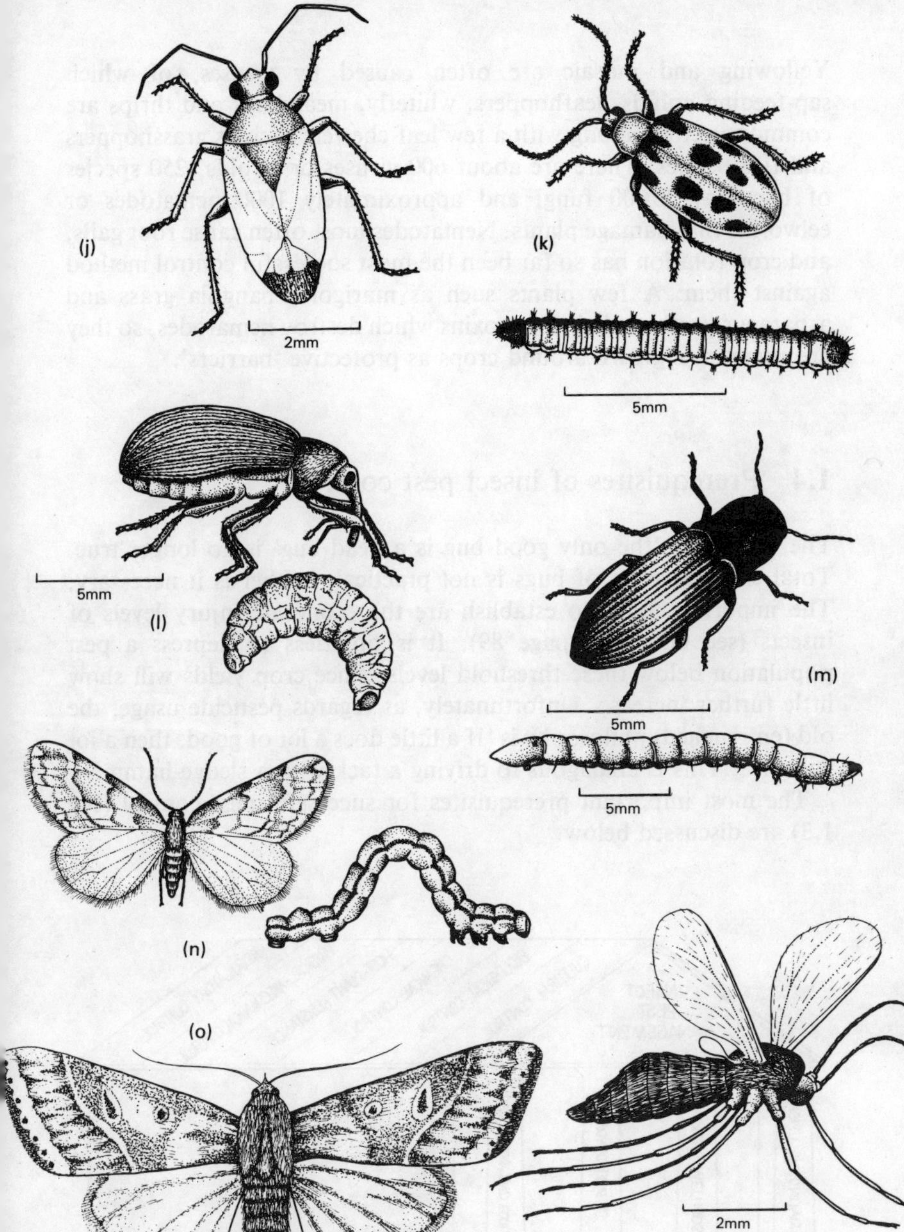
(j)
2mm
(k)
5mm
5mm
(l)
(m)
5mm
5mm
(n)
(o)
5mm
2mm
(p)

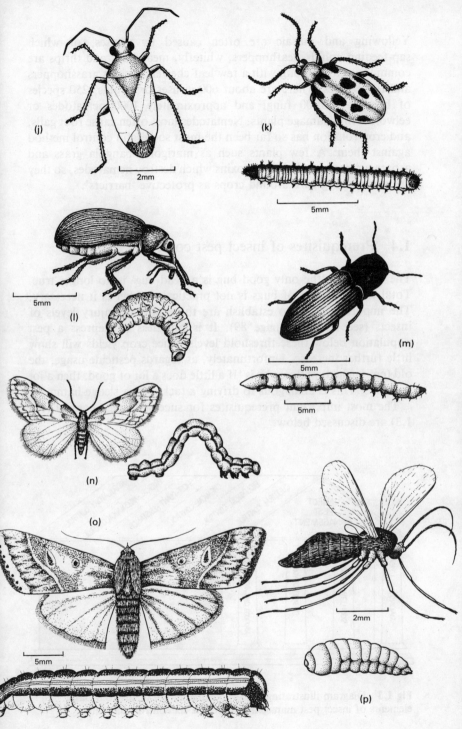

Yellowing and mosaic are often caused by viruses, of which sap-feeding aphids, leafhoppers, whitefly, mealybugs and thrips are common vectors, along with a few leaf chewers such as grasshoppers and flea-beetles. There are about 600 viruses or viroids, 250 species of bacteria, 20 000 fungi and approximately 1000 nematodes or eelworms that damage plants. Nematodes most often cause root galls, and crop rotation has so far been the most successful control method against them. A few plants such as marigold, pangola grass and asparagus produce their own toxins which destroy nematodes, so they are sometimes grown around crops as protective 'barriers'.

1.4 Prerequisites of insect pest control

The old adage 'the only good bug is a dead bug' is no longer true. Total extermination of bugs is not practical, neither is it necessary. The important facts to establish are the economic injury levels of insects (see Glossary, page 89). It is pointless to depress a pest population below these threshold levels, since crop yields will show little further increase. Unfortunately, as regards pesticide usage, the old (entrenched) philosophy is 'If a little does a lot of good, then a lot will . . .'. This is analagous to driving a tack with a sledge-hammer.

The most important prerequisites for successful pest control (Fig 1.3) are discussed below.

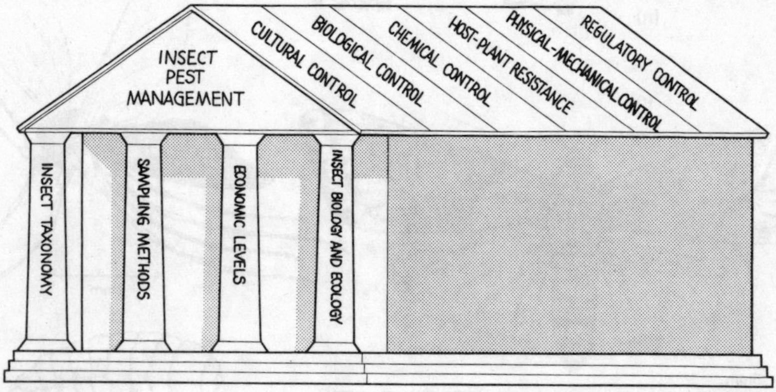

Fig 1.3 Diagram illustrating the supporting nature of the four basic elements of insect pest management and the six control techniques.

1 A knowledge of insect taxonomy

In the US, misidentification of the beet leafhopper as *Eutettex tenellus* led to searches for natural enemies in South America, its supposed home. Only after reidentification, (to a different genus!—*Circulifer*), were successful searches performed in the Mediterranean area. In 1972 an outbreak of 'cotton budworms' in Arizona could not be controlled by insecticides. Only after the pest was found to be tobacco budworm was the outbreak controlled by higher doses.

2 Good sampling techniques for the establishment of insect pest life-histories

Sampling is usually performed once or twice a week. Methods vary according to whether eggs, larvae, pupae or adults are sampled. Common techniques involve sweep netting, random or point counting of pests per plant, light trapping, sticky or pheromone trapping and sometimes estimating feeding damage.

3 A knowledge of economic thresholds

Complete pest elimination is pointless: the most cost-effective management technique is to keep populations only just below economic thresholds. For example, sugar beet can withstand 50% defoliation by the mangold fly, *Pegomyia hyoscyami*, with no loss in the percentage of sugar in the beet and only a 5% drop in yield. In cereals, the lower leaves contribute only 15–20% to grain yield so some damage to these leaves is acceptable. Market prices, consumer needs or preferences, and federal or state regulations may also determine economic damage thresholds. Besides determing economic thresholds, many growers also apply control measures in an economically optimal way, according to cost-benefit analyses similar to those outlined in Fig 1.4.

4 A knowledge of pest biology, ecology and natural enemies

For successful pest control, a knowledge of how natural population controls operate is vital. This entails studies of natural enemies (predators and parasites) and the influences of abiotic variables.

Having established these corner-stones of pest management, we can begin to manipulate population densities using different control techniques. Before examining present control methods, however, it is worthwhile to examine the history behind modern concepts.

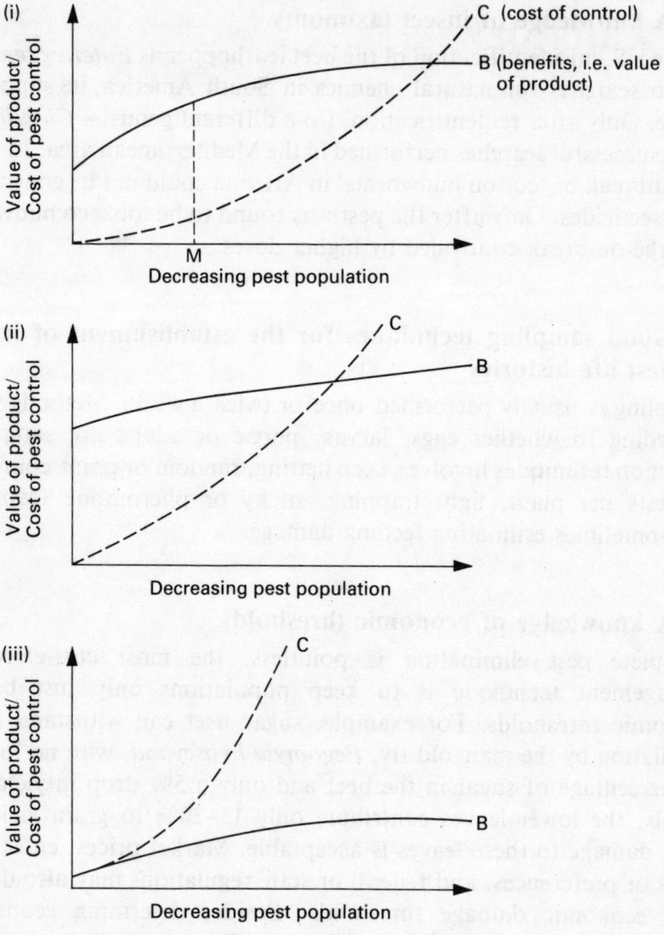

Fig 1.4 Hypothetical cost-benefit analyses of individual pest problems. (i) Optimal control strategy (i.e., largest value of B–C) is at M. (ii) Optimal pest control (largest value of B–C) occurs when cost of control is zero. (iii) Pest population is below economic levels. Cost of control always exceeds benefits, so pest control action cannot be justified.

2 A history of pest management

The transition from the hunter-gatherer way of life to primitive agriculture, some 10 000 years ago, signalled the rise to prominence of many 'pest' species. Agricultural societies are thought to have evolved independently in at least three locations: South China (rice); the Near East, modern Iran and Iraq (wheat) and Central America (maize). Well before 2500 BC the Sumerians were using sulphur compounds to control insects and mites. By 1200 BC botantical insecticides (naturally occurring insecticides derived from plants) had been used for seed treatments and as fungicides in China. The Chinese were very advanced and used mercury and arsenic compounds to control body lice. It was in China in about 300 AD that biological control was first established. Colonies of predatory ants, *Oecophylla smaragdina*, were set-up in citrus groves to control caterpillars and beetles. Bamboo bridges were built to facilitate movement of the ants from tree to tree. Alteration of planting dates for crops, so as to avoid insect injury, was practised in China, several centuries before Christ. Ko Hung, the greatest alchemist of the 4th century, recommended a root application of white arsenic when transplanting rice, to protect it against insect pests.

The great Roman Empire was battling with pests too. In 200 BC Cato the Censor advocated oil sprays for pest control and in 13 BC the first pest-proof (rat-proof) grannary was constructed by the architect Marcus Pollio. Not all pest control practices in Rome were so well-founded biologically. Faced with the almost insurmountable problems of locust control, the texts recommended: 'A woman, ungirdled, with flying hair, must run barefoot around the garden.' Alternatively: 'A crayfish must be nailed up in different places in the garden.'

A complicated charm against damage to millet by sparrows and worms was as follows: before sowing, carry a toad around the field at night, place it in a pot and bury it in the middle of a field; dig-up the pot before broadcasting the seed, otherwise the earth will become sour. Presumably the venom of the toad frightened birds and maggots away by magic—but it could not be left too long or the toad would frighten the soil too!

Whilst pest control in certain parts of the world flourished, other regions appeared to be making little headway. In Berne, Switzerland,

Table 2.1 Major events in the history of pest control

Date	Event
400 000 000 BC	First land plants
350 000 000 BC	First insects
250 000 BC	Appearance of *Homo sapiens*
12 000 BC	First records of insects in human society
8000 BC	Beginings of agriculture
4700 BC	Silkworm culture in China
2500 BC	First records of insecticides
1500 BC	First descriptions of insect pests
950 BC	First descriptions of cultural controls (burning)
300 AD	First record of use of biological controls (predatory ants used in citrus orchards in China)
1650–1780	Burgeoning of insect descriptions (after Linnaeus) and biological discoveries in Renaissance
1732	Farmers begin to grow crops in rows to facilitate weed removal
1750–1880	Agricultural revolution in Europe
Early 1800s	Appearance of first books and papers devoted entirely to pest control
1840s	Potato blight in Ireland (no controls available to curb disaster)
1870–1890	Grape Phylloxera and powdery mildew controlled in French vineyards (by the introduction of Bordeaux mixture and Paris Green and the use of resistant rootstalks and grafting)
1880	First commercial pesticide spraying machine
1888	First major success with an imported biological control agent (vedalia beetle for control of cottony-cushion scale)
1890s	Introduction of lead arsenate for insect control
1896	Recognition of arthropods as vectors of human disease
1896	First selective herbicide (iron sulphate)
1901	First successful biological control of a weed (lantana in Hawaii)

(continued)

Table 2.1 (Continued)

Date	Event
1899–1909	Development of strains of cotton, cowpeas and water-melon resistant to *Fusarium* wilt (first breeding programme)
1915	Control of disease-vector mosquitoes allowing completion of Panama Canal
1921	First aircraft spray (against Catalpa sphinx moth in Ohio)
1929	First area-wide eradication of an insect pest (against Mediterranean fruit fly in Florida)
1930	Introduction of synthetic organic compounds for plant pathogen control
1939	Recognition of insecticidal properties of DDT
1940	Use of milky disease to control Japanese beetle (first successful use of insect pathogen for control)
1940	Organophospates developed in Germany, carbamates in Switzerland
1942	First successful breeding programme for insect pest resistance in crop plants (release of wheat strain resistant to hessian fly)
1944	First hormone-based herbicide (2,4–D)
1946	First report of insect resistance to DDT (housefly in Sweden)
1950s, 1960s and 1970s	Widespread development of resistance to DDT and other pesticides
1950s	First applications of systems analysis to crop pest control
1959	Introduction of concepts of economic thresholds, economic levels and integrated control
1960	First insect sex pheromone isolated, identified and synthesised (gypsy moth)
1970s	Banning of DDT

as late as 1476 AD cutworms were taken to court, pronounced guilty, excommunicated by the Archbishop and banished. In 1485 the High Vicar of Valence commanded caterpillars to appear before him, gave them a defence council and, finally, condemned them to leave the area.

During the Renaissance ('rebirth of knowledge', 1400–1600) in Europe, more emphasis was at last placed on scientific knowledge. With the development of the microscope, Van Leeuwenhoek discovered bacteria in 1675 and Reni proved that insects do not arise spontaneously from decaying matter but from eggs laid there. In the first half of the 18th century Carl Von Linne (Linnaeus) laid the foundations of modern systematic nomenclature (taxonomy) which provided a firm basis for pest identification. Linnaeus was also interested in pest control and in 1763 won a prize for an essay, under the name of C. N. Nelin, on how orchards could be freed from caterpillars. He suggested mechanical, chemical and biological control methods. The biological methods involved the predatory beetle, *Calasoma sycophanta*, 'like a wolf among the sheep, creating havoc among the caterpillars.'.

Reamur (1683–1756) discussed the significance of host-parasite relationships in pest outbreaks and suggested the use of entomophagous (insect-eating) insects, especially lacewings, as control agents against aphids in greenhouses. Another famous entomologist of the time, Latrielle (1762–1833), fell foul of the revolution in France. Whilst awaiting the guillotine he persuaded the prison doctor to convey a specimen of a 'very rare' insect that he had found to friends in town. This was his only way of letting them know where he was. The friends remonstrated with the authorities, Latrielle was freed, and later the name of the insect, *Necrobia ruficollis*, which is in fact quite common, was engraved on his tombstone.

The period 1750–1880 was also a time of agricultural revolution in Europe. Crop plantations became more extensive and international trade promoted the discovery of the botanical insecticides pyrethrum and derris. Unfortunately, international trade and travel also promoted pest travel between continents, resulting in some of the worst pest outbreaks ever known. The potato blight of Ireland, England and Belgium occurred in the 1840s and fungus leaf spot appeared on coffee in Ceylon, causing a switch to the production of tea. The invasion of Europe by the American insect, grape Phylloxera, *Viteus vitifoliae*, nearly put an end to the French wine industry between 1848 and 1878. The dangers of toxic poisons also became known around this time when, in 1754, Aucante of France observed arsenic poisoning of field workers. In 1786 the use of arsenic and mercury for seed treatments was prohibited in France.

By the early 1840s books devoted entirely to pest control had appeared. They covered most of the essential elements which are still found in today's methods: cultural control (adjustment of planting time, and fertilisation); biological control (encouraging natural enemies); varietal control (selecting pest-resistant varieties); physical control (construction of physical barriers) and chemical control using a variety of substances. Materials for chemical control of pests did not change much in the 50 years after 1880, the active ingredients still usually containing arsenic, antimony, selenium, thallium, zinc, copper or plant-derived alkaloids. Chemical control of weeds found its first application in 1896 when iron sulphate was found to kill broad-leaved weeds but not crops. However, labour was so cheap at this time that most growers relied on hand-weeding.

It was also in the late 19th century that the importation and establishment of natural enemies for biological control was shown to be effective. The first major success using this technique was control of the cottony cushion scale, *Icerya purchasi*, which had been accidentally introduced into California in the 1860s and by the 1880s had nearly wiped-out the citrus industry. The native home of this pest was found to be Australia, so in 1877 the US government sent an entomologist, Albert Koebele, to Australia to find natural enemies. Koebele sent back two: a parasitic fly, *Cryptochaetum iceryae*, and the vedalia beetle (a ladybird beetle), *Rodolia cardinalis*. The vedalia beetle provided very effective control. Later 140 of these predators were carefully shipped back to California and within 18 months their descendents had checked the growth of cottony cushion scale over the citrus growing areas of the entire state. Koebele returned to his native Germany just before the 1914–1918 war and never succeeded in gaining readmission to the USA despite his huge service to that country. After 1890 the vedalia beetle was effective enough to control the cottony cushion scale single-handedly. In the 1950s DDT killed the vedalia beetle allowing the scale to reach economically injurious levels once more.

It was during the 1890s that arthropods were proved to be disease vectors (Table 2.2). Methods of controlling diseases by removing the

Table 2.2 The years in which some disease vectors were discovered

Year	Disease	Vector
1893	Texas cattle fever	Ticks
1896	African sleeping sickness	Tsetse flies
1896	Plague or 'black death'	Rat fleas
1897	Malaria	Mosquitoes
1898	Typhoid	Flies
1900	Yellow fever	Mosquitoes

vector began to be implemented. The Panama Canal, abandoned by the French in the late 1800s, because of the effects of malaria and yellow fever, could be completed in 1915.

In the 1940s the new insecticides made things all too easy for growers. Pest control, a fundamentally ecological problem, began to assume the nature of an offshoot of chemistry and engineering. This trend was reflected in publications of the *Journal of Economic Entomology*. After 1935 the number of papers investigating general biology decreased and field-testing of insecticides began to dominate the journal (Fig 2.1). Disaster had struck when the first reported case of DDT resistance, in Swedish house-flies, appeared in 1946. Within 20 years some 224 insect species and acarids were recorded as resistant to one or more groups of insecticides. Of these, 127 were of agricultural importance and 97 were of medical or veterinary importance. By 1975, 75% of the most serious pests in California had developed resistance to at least one major insecticide.

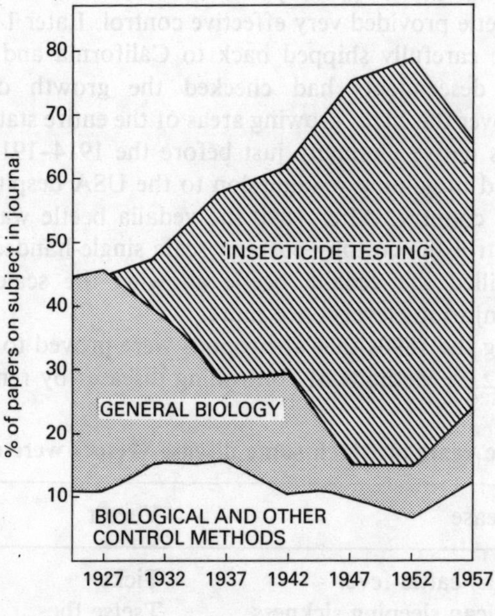

Fig 2.1 Trends in applied entomological research as reflected by papers in the *Journal of Economic Entomology*, 1927–1957.
(Reproduced, with permission, from *The History of Entomology*, © 1973 by Annual Reviews Inc.)

3 Insect morphology, development and classification

Before dealing with control techniques it is valuable to examine insect anatomy and development so that the vulnerable periods in an insects life history can be pinpointed and control measures can be implemented when they will be most effective.

3.1 Morphology

All insects are covered by an outer skeleton of hard or soft chitinous cuticle. This is colourless and transparent in its pure state (as in the wings of a fly) but often occurs 'tanned', or sclerotised, incorporating strengthening elements. Chitin is a nitrogenous polysaccharide and exhibits properties similar to cellulose, the chief ingredient of cotton and nylon. The tough exoskeleton supports the internal organs, prevents excessive water loss and is also resistant to the action of some of the strongest chemicals. Periodically, during the immature stages, insects wriggle out of their chitinous cuticles ('moult') and form new ones. Some immature stages have only a thin, somewhat elastic cuticle for their entire development. Such larvae appear to be ideal candidates for control by contact chemicals. However, many are secretive, living inside plant tissue or in the soil, both regions being inaccessible to sprays.

The insect body, well represented by the grasshopper and the beetle (Fig 3.1) is divided into three well-defined regions: the head; thorax and abdomen. The head bears most of the sensory organs, the thorax bears the legs and wings and the abdomen contains the reproductive system and much of the digestive tract.

3.1.1 The head

Every insect has a pair of antennae located on its head. These vary greatly in length, shape and number of segments but all are sensory in function. Where long and flexible, as in grasshoppers, antennae function as sensitive feelers. In flies they have been found to possess the sense of smell, whilst in male mosquitoes they serve an auditory function. They may be used in communication as in ants, to smell females, as in moths, or to find food.

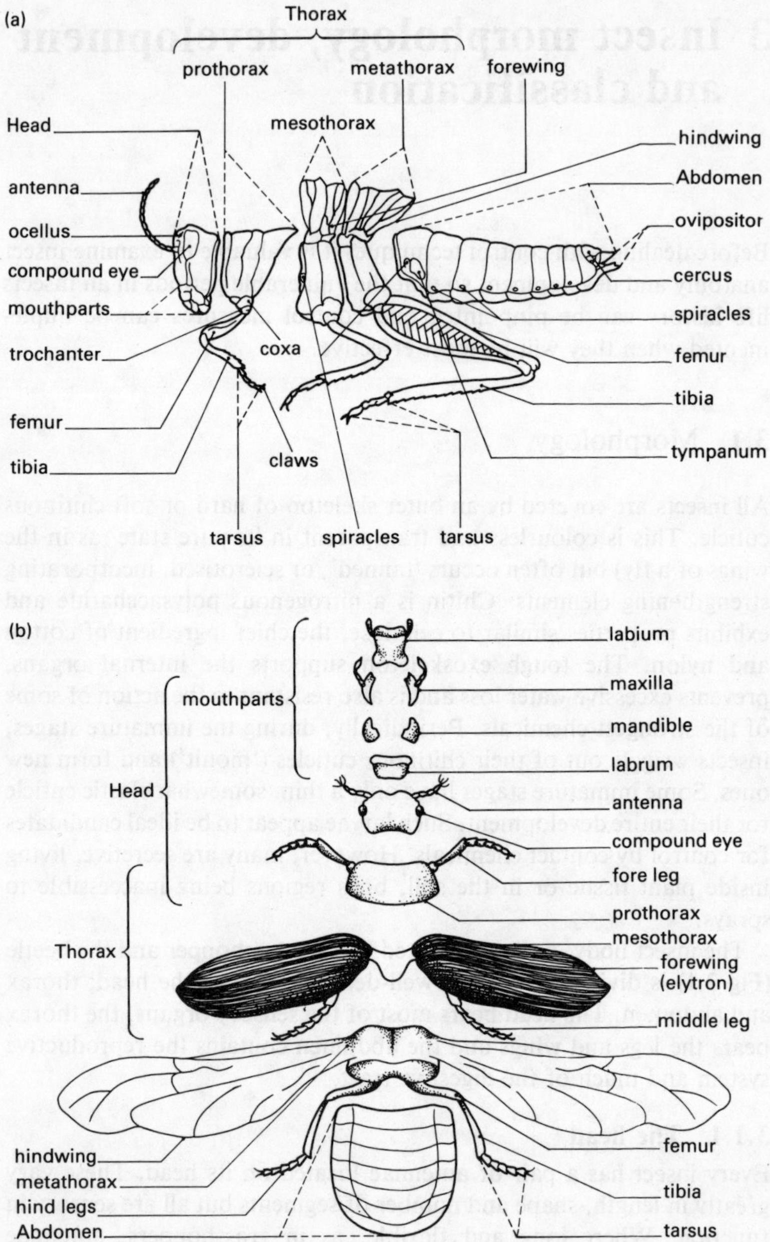

Fig 3.1

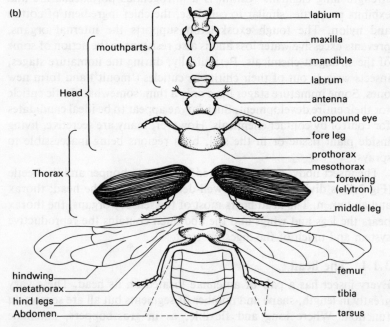

(a) An exploded drawing of a grasshopper showing the main body regions and appendages.

(b) An exploded drawing of a beetle showing its external structures.

(Page 3, Figs 2 and 3, Fundamentals of Entomology and Plant Pathology, 2nd edition by Pyenson, AVI Publishing Company, Inc., Westport, Conn., 1980.)

Insect eyes are of two basic types: simple eyes (ocelli) and compound eyes. All larvae possess ocelli which are bubble-like visual units considered to function in light response activities only, not as organs of acute vision. Adult insects commonly possess a pair of compound eyes in addition to simple ocelli. Each compound eye is composed of hundreds of individual facets each of which gives a visual image so that together the overall picture is rather like a mosaic. Compound eyes are usually bulbous organs situated on top of the head, so an insect's field of vision is practically 360°, a valuable asset in evading natural enemies.

Insect mouthparts vary greatly in structure and function, depending of feeding habits (Fig 3.2). The basic, or primitive, kind is the chewing type which allows plant parts to be bitten off or tunnels to be made into plant or animal tissue. Chewing mouthparts consist of a labrum, or upper lip, which helps cover the mouthparts from above, two heavy chitinised grinding organs called the mandibles, which work on a lateral plane biting off the food, two slimmer chitinised organs called the maxillae, which work in unison with the mandibles, and a flap-like under lip, called the labium. Associated with these mouthparts are several more 'fleshy' sensory organs, called palps, which taste or smell the food. These are located on the maxillae (maxillory palps) and the labium (labial palps and the hypopharynx). Many insect larvae as well as adult beetles and grasshoppers have these biting mouthparts. Biting insects lend themselves to control by stomach poisons, the poison being sprayed on foliage or fruit and ingested with the food.

Piercing and sucking mouthparts are thought to have evolved from the chewing type by the elongation of the labium and its formation into a tube. In a similar manner the labrum has become elongated into a slim v-shaped structure and the two mandibles and maxillae have become hair-like and sharpened for piercing. These are now given the name 'stylets' and are enclosed in the protective beak formed by the labium and labrum. In some species, the maxillary and labial palps have been lost so that the stylets fit closely together. These piercing and sucking mouthparts are found in all the hemiptera, aphids, scales and leafhoppers as well as blood-sucking fleas, flies and lice. Although piercing mouthparts may leave no visible scar when withdrawn, plant and animal diseases are often injected with the saliva directly into the sap or bloodstream. Obviously, stomach poisons sprayed onto foliage have little effect, so these pests are best controlled using systemics injected into the sap or by contact poisons.

There are several modifications of the piercing and sucking mouthpart design. Thrips (Thysanoptera) possess three stylets in a short conical beak that rasp away plant tissue so that juice exudes.

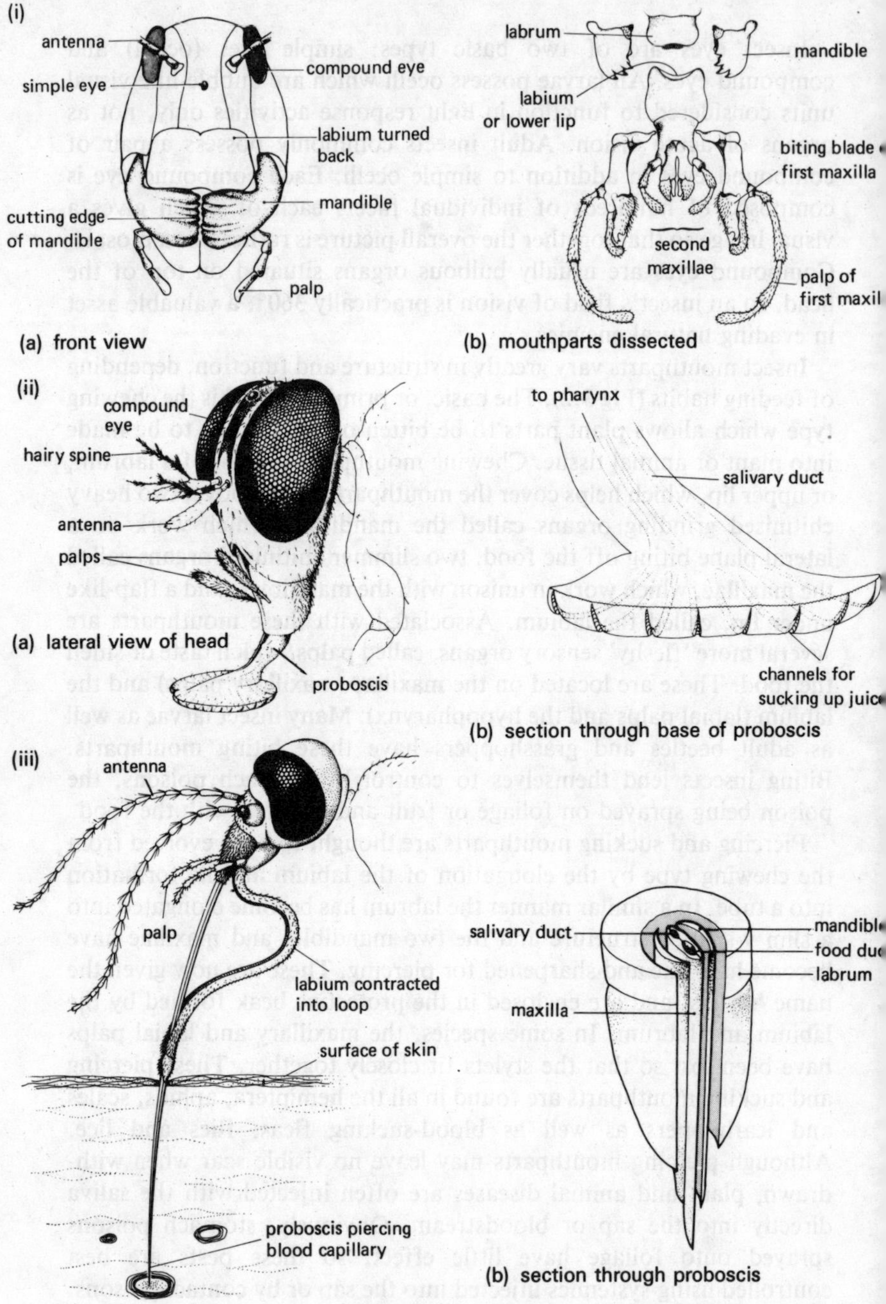

Fig 3.2 Insect mouthparts: (i) biting mouthparts (grasshopper); (ii) sucking mouthparts (housefly); (iii) piercing and sucking mouthparts (mosquito).

This is then sucked up via the conical beak. Another modification is exhibited by certain flies, including the housefly, which possess a fleshy proboscis with a sponge-like organ at the end for ingesting liquids. The proboscis of the Lepidoptera is no more than a long tube, normally coiled under the head and extended to suck nectar from flowers. Bees have retained sharp chewing mandibles. They also possess modified maxillae and a labium designed to lap up plant nectar.

3.1.2 The thorax

Situated immediately behind the head of an insect is the thorax. Typically it is divided into three distinct segments, each of which bears a pair of legs. The first thoracic segment, the prothorax, never bears wings, whilst, in most adult insects, the second and third segments, the mesothorax and metathorax, each have one pair. Wings are cuticlar outgrowths from the body and they usually bear thickened veins. Venation is constant for a species and is a prime taxonomic aid. In some instances, particularly in the beetles, the forewings lose their flying function and become protective covers, called elytra, for the hindwings. In true flies, the Diptera, there is only one pair of wings, the second pair being represented by slender balancing organs called halteres.

Insect legs are all generally alike, consisting of a coxa (which fits into the body like a ball and socket joint) a small and inconspicuous trochanter (like a heavy femur) a long and slender tibia and a tarsus (a foot-like appendage of five segments). In some species, legs may be modified to perform specific functions. The enlarged hind legs of grasshoppers are modified for jumping, the front legs of mole crickets for digging and the front legs of preying mantises for catching and holding prey.

3.1.3 The abdomen

The segments that make-up the abdomen of insects are generally similar in form. The abdomen tapers towards the posterior where the segments are often modified. The paired small openings on the lateral sides of each segment, save the last two, are known as spiracles. It is through these that air diffuses into the body via a network of branching tracheae. Each cell in the body receives its air supply through a very small trachea, or tracheole. It is through the spiracles that many insecticides enter the body. Some insects can regulate their air intake by opening and closing the spiracles. This is thought to be one of the causes of the variation in suseptibility of insects to toxic gases.

True legs are never present on an insects abdomen, but in some immature stages, such as caterpillars, paired unsegmented appendages, called prolegs, are present. These are used in clinging to surfaces. At the end of the adult female abdomen there is often an ovipositor, used for guiding and depositing eggs. The length and shape varies according to the substrate in which the eggs are laid; if eggs are simply deposited on plant surfaces, the ovipositor may be totally absent. In the social insects where only the queen lays eggs, the ovipositors of other females are modified into stings which serve in defence or to paralyse prey.

3.2 Classification

Combinations of these anatomical characters are used in insect classification, similar insects being grouped together in similar taxa. The smallest commonly used taxonomic category is the species, which may be defined as a reproductively isolated, naturally distinct group of organisms. Closely-related species are grouped together in a genus. Whilst many insects have easily recognisable common names, such as the cabbage butterfly, these names often differ between countries or even regions. For standardisation, a scientific name is assigned, so that each species has its own, universal name. The scientific name consists of the genus or generic name combined with the species or specific name. For the cabbage butterfly this is *Pieris rapae*. This classification system was established by the Swedish naturalist, Carl von Linne, in 1758 and has been widely adopted ever since. Genera of similar characteristics are combined to form families, similar familes form an order, similar orders a class and similar classes a phylum (animals) or a division (plants). The phyla are combined into the animal kingdom; the plant divisions make up the plant kingdom. These two kingdoms include nearly all living organisms known on earth. Two examples of insect classification are given below.

Common name	**Codling moth**	**Squash bug**
Kingdom	Animal	Animal
Phylum	Arthropod	Arthropod
Class	Insecta	Insecta
Order	Lepidoptera	Hemiptera
Family	Olethreutidae	Coreidae
Genus	*Laspeyresia*	*Anasa*
Species	*pomonella*	*tristis*
Scientific name	*Laspeyresia pomonella*	*Anasa tristis*

The class Insecta is divided into 26 orders on the basis of characteristics such as presence or absence of wings, wing venation, mouthpart structure, antennal structure and distribution of leg and body bristles. A simple key to the main orders is illustrated in Fig 3.3.

3.3 Insect development

There are probably more misconceptions about stages in insect life histories than about those of higher animals. This is partly due to their bewildering variety of developmental methods.

3.3.1 Insect eggs

The life of each insect may be said to begin with the formation of an egg in the female's ovary. Although fertilisation by the male is usually necessary to permit further development, some insects are capable of producing living young without this – a process known as parthenogenesis. As may be expected, some important pests, including aphids and some beetles and thrips, reproduce extremely quickly by this method. Fortunately, many insect parasites can also reproduce parthenogenetically and keep pace with outbreaks of such pests.

Insect eggs are often overlooked, being small and camouflaged or concealed. It is often vital to identify the exact kind of insect that will develop from an egg, since many pests can be controlled at this stage, before economic damage is done by the larvae or adults. Egg taxonomy involves such features as shape, size, colour, chorion sculpturing and method of placement in or on host tissue. Batch size can also be an important factor: the large, clumped egg-masses of certain moths, for example, lend themselves more readily to control by parasites than single well concealed eggs.

Although most insects are oviporous (that is, they lay eggs from which the young hatch) some, notably aphids and some scale insects, are ovoviviporous (that is, the eggs are retained within the female's body until they have hatched and live young are produced). Viviparous reproduction (as found in man and other mammals) is even rarer, but is nevertheless present in one or two species such as the tsetse fly and sheep tick. Here the embryos develop in the body cavity and are fully mature when born. The potential for biological control in these species is severely reduced.

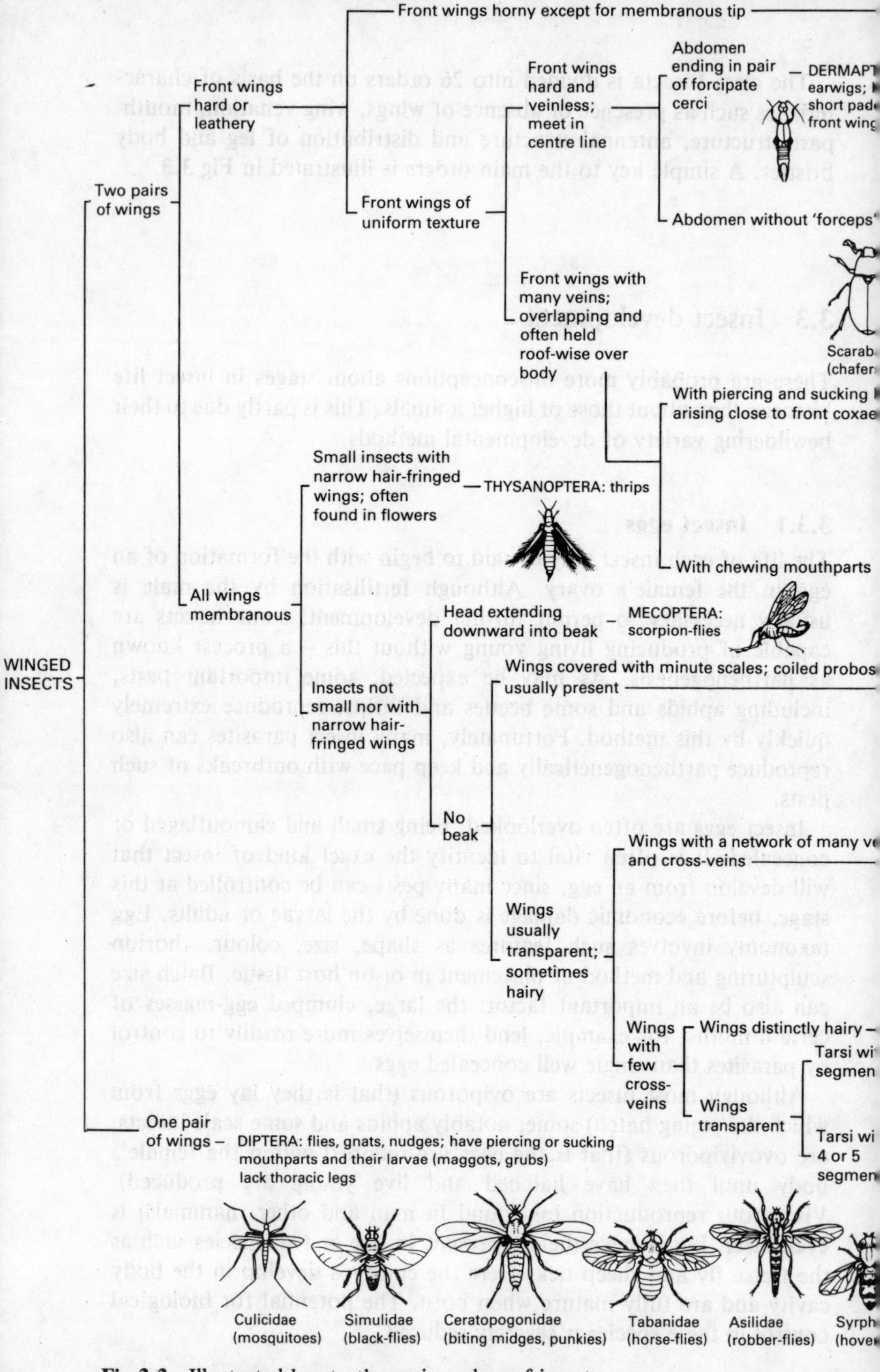

Fig 3.3 Illustrated key to the main orders of insects.

EROPTERA (Hemiptera): true bugs; have wings folded flat on abdomen and a piercing 'beak'

Reduvidae (assasin bugs) | Coreidae (squash bugs) | Lygaeidae (chinch bugs) | Miridae (Plant bugs) | Pentatomidae (stink bugs)

OLEOPTERA: beetles; have biting mouthparts and larvae (grubs) with or without true legs

prionidae (...ing beetles) | Coccinellidae (ladybird beetles) | Chrysomelidae (leaf beetles) | Cerambycidae (long-horned beetles) | Curculionidae (weevils) | Scolytinae (bark beetles) | Bruchidae (seed beetles)

OMOPTERA (Hemiptera):

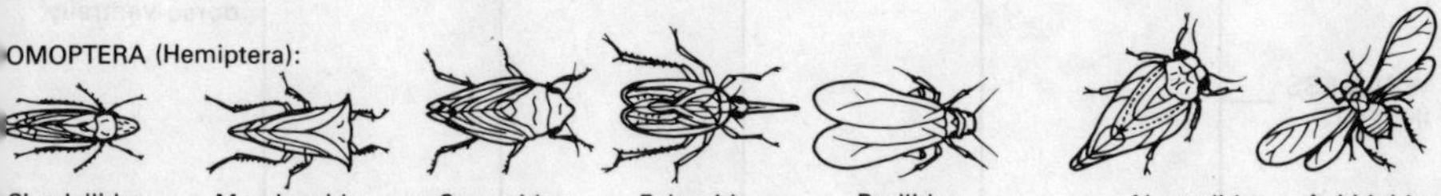

Cicadellidae (leaf hoppers) | Membracidae (tree hoppers) | Cercopidae (spittle bugs) | Fulgoridae (plant hoppers) | Psyllidae (jumping plant-lice) | Aleyrodidae (white-flies) | Aphidoidea (aphids)

nd legs modified for jumping – ORTHOPTERA: grasshoppers, crickets, katydids; have biting mouthparts, tympanic organs and long antennae

Tettigoniidae (grasshoppers)

nd legs not modified for jumping – DICTYOPTERA: cockroaches, mantids

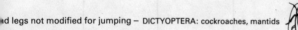

cockroach

PIDOPTERA: moths and butterflies; have phytophagous larvae (leaf chewers, leaf miners or stem borers) with true thoracic legs, and prolegs on abdomen

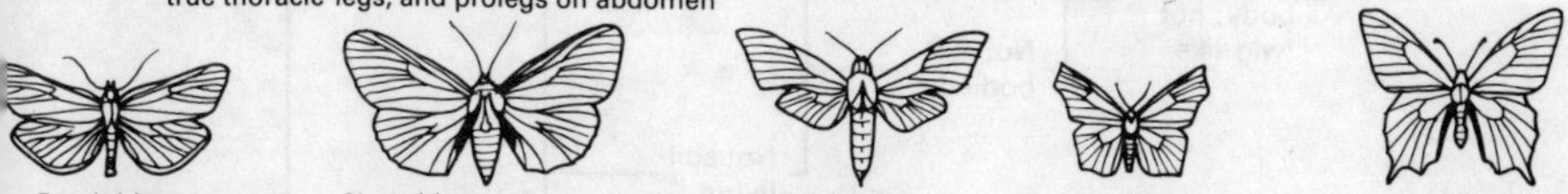

Pyraloidea (snout moths) | Noctuidae (cutworms, owlet moths) | Sphingidae (hawk moths) | Geometridae (inch-worms, geometers) | Papilionoidea (butterflies)

HEMEROPTERA, ODONATA, PLECOPTERA and NEUROPTERA

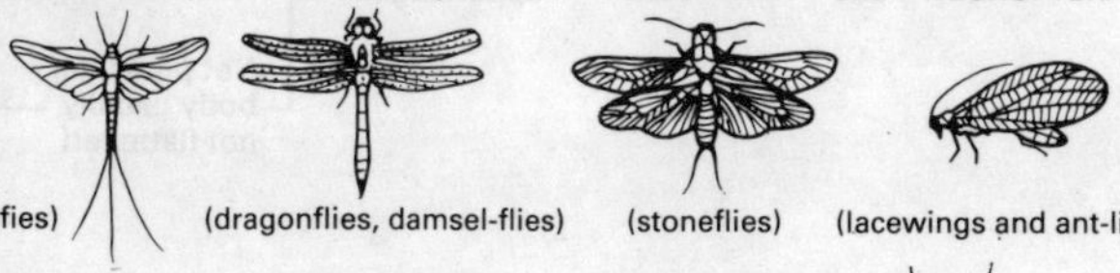

yfies) | (dragonflies, damsel-flies) | (stoneflies) | (lacewings and ant-lion flies)

CHOPTERA: caddis-flies; have aquatic, often case-building larvae.

OCOPTERA: booklice; tiny insects with at least 12 antennal segments

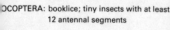

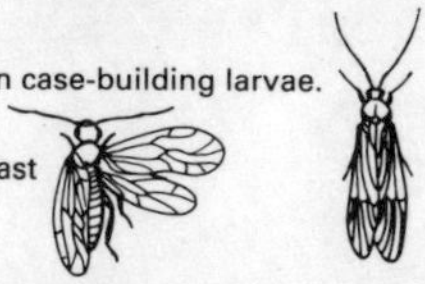

wings alike —— ISOPTERA: termites; social insects, with sterile workers and soldiers

nd wings smaller than front wings – HYMENOPTERA: bees and wasps

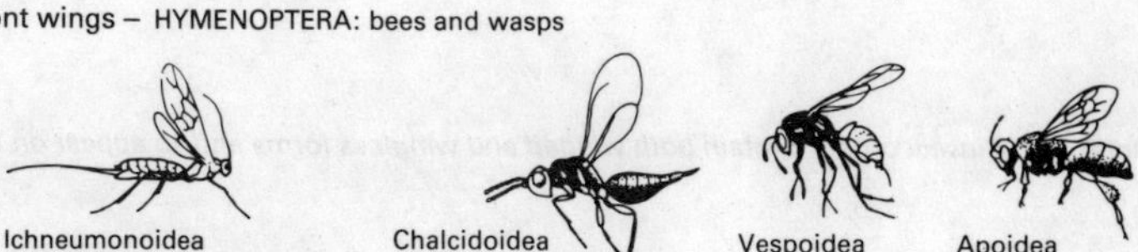

Ichneumonoidea (ichneumons and braconids; important insect parasitoids) | Chalcidoidea (chalcid wasps; important parasitoids) | Vespoidea (social wasps) | Apoidea (bees; important plant pollinators)

WINGLESS INSECTS

- Long, slender body; twig-like – PHASMIDA: stick insects 
- Do not have long slender body; not twig-like
  - Small, soft-bodied with protective shield or scale – HEMIPTERA: scale insect 
  - Not soft-bodied
    - Minute; live mainly in soil – PROTURA, DIPLURA and COLLEMBOLA
    - Not soil-living
      - With abdominal appendages – THYSANURA: bristle
      - No abdominal appendages
        - Parasitic; generally flattened laterally dorso-ventrally
        - Not parasitic; body usually not flattened

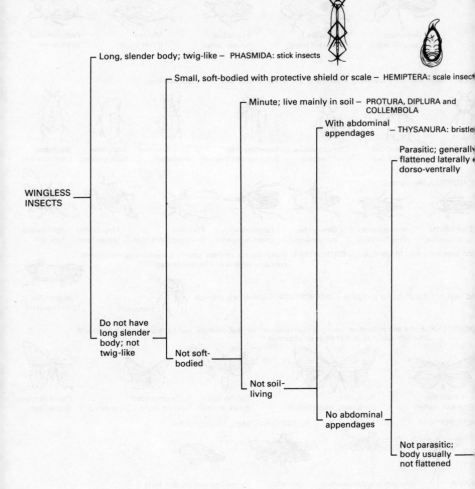

Note: *many insect orders contain both winged and wingless forms and so appear on both keys.*

Fig 3.3 (continued)

- Jump; flattened laterally – SIPHONAPTERA : fleas

- Do not jump; flattened dorso-ventrally
  - Head partly withdrawn into thorax
    - Short antennae; strong claws – DIPTERA: true flies

    - Longer antennae; claws less prominent – HEMIPTERA: bugs

Cimicidae (bed-bug)

  - Head not withdrawn into thorax
    - Biting mouthparts; a distinct prothorax – MALLOPHAGA: biting lice

    - Sucking mouthparts; thoracic segments fused – ANOPLURA: sucking lice

- Abdomen constricted to a 'waist' —— HYMENOPTERA: ants

Formicidae (ants)

- No 'waist'
  - Body clothed with flattened hairs and scales; wing vestiges present —— LEPIDOPTERA
  - Body naked
    - Antennae long and slender; frequently found in dry materials — PSOCOPTERA: booklice

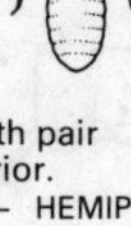

    - Antennae shorter; abdomen often with pair of tubular outgrowths towards posterior. Commonly found on growing plants – HEMIPTERA: aphids

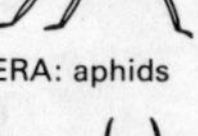

    - Antennae short and bead-like; often lacking pigment (whitish) – ISOPTERA: termites

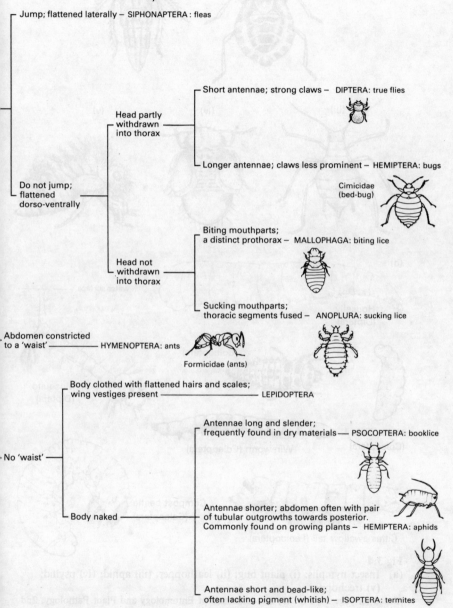

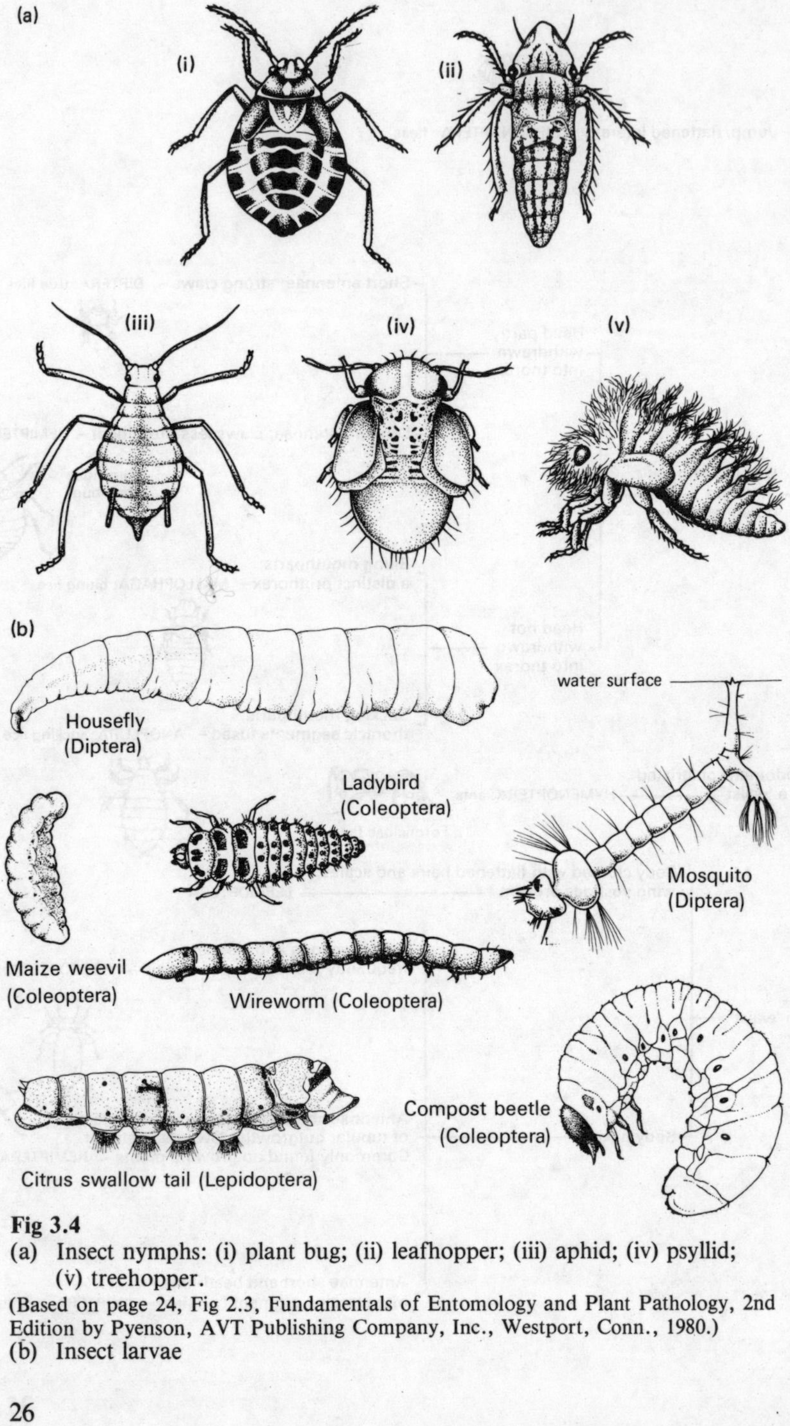

Fig 3.4

(a) Insect nymphs: (i) plant bug; (ii) leafhopper; (iii) aphid; (iv) psyllid; (v) treehopper.

(Based on page 24, Fig 2.3, Fundamentals of Entomology and Plant Pathology, 2nd Edition by Pyenson, AVT Publishing Company, Inc., Westport, Conn., 1980.)

(b) Insect larvae

3.3.2 Insect immature stages

There are two basic methods of development from egg to adult in insects. In one method, newly-hatched insects are like miniature adults in appearance except that they lack wings (Fig 3.4). These young are termed nymphs and develop into adults via about five moults, each successive instar being larger. The wings also develop gradually from small wing buds in young nymphs to the fully developed adult wings. Insects with this type of gradual development generally have piercing mouthparts, as in the case of Homoptera, such as leafhoppers, thrips and scales. More rarely, they have biting mouthparts, as in Orthoptera, the grasshopper and crickets.

Table 3.1 Identification of agriculturally important larvae

A With thoracic legs but without abdominal prolegs.
- 1 Head distinct, sometimes depressed, mouth directed forward or downward, no adfrontal area. Larvae of **Coleoptera** (beetles).
- 2 Head usually globular, adfrontal area usually present, mouth directed downward, spinneret present near end of labium. Larvae of **Lepidoptera** (moths, butterflies).

B With thoracic legs and abdominal prolegs.
- 1 Two, three, or five pairs of prolegs with tiny hooks (crochets), more than one pair of simple eyes of none on each side of head. Larvae of **Lepidoptera**.
- 2 Prolegs, usually six to eight pairs, no tiny hooks, one simple eye on each side of head or none. Larvae of **Hymenoptera** (sawflies).

C No thoracic legs, no prolegs, distinct head.
- 1 Labium with a projecting median spinneret. Larvae of **Lepidoptera**.
- 2 Body generally short, frequently U-shaped, pair of spiracles on each of the principle abdominal segments. Larvae of **Coleoptera** (curculios).

D No thoracic legs, indistinct head.
- 1 Abdomen with several pairs of spiracles or none, no antennae, body tapers somewhat at both ends. Soft, white or yellow grubs found in wax, in paper cells in bodies of insects or in galls. Larvae of **Hymenoptera** (bees, wasps, ants).
- 2 Abdomen usually with one pair of spiracles located on blunt end. Head end pointed. Mouthparts usually consisting of a pair of hooks. Larvae of **Diptera** (flies).

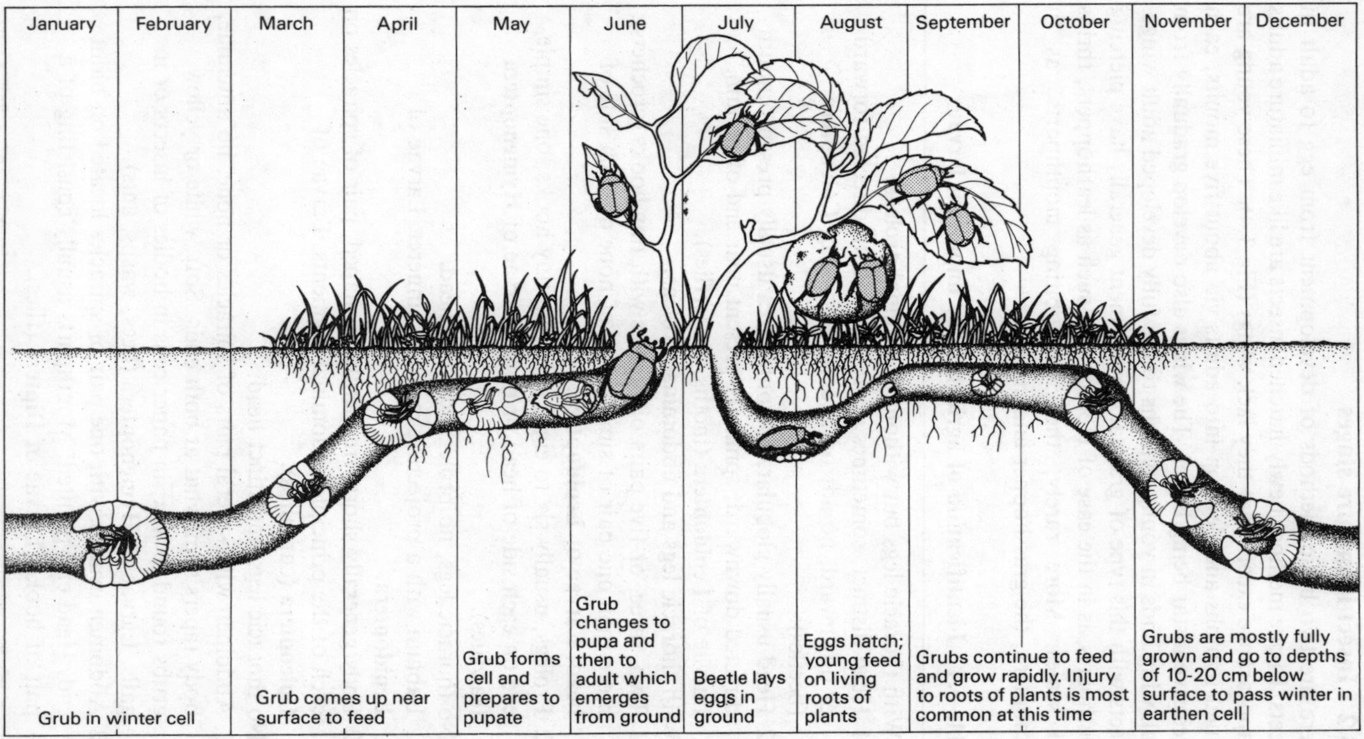

Fig 3.5 The life history of the Japanese beetle.

In contrast to nymphs, many insect immatures are worm-like and are known as larvae. Larvae show no traces of wings during any period of growth and often differ greatly in habits from the adult (Fig 3.5). Larvae also undergo moults and pass through several instars before they reach the size characteristic of their species. Larvae have simple eyes and sometimes bear additional appendages, known as prolegs, as locomotory aids. Their mouthparts can be radically different from those of the adult owing to a different diet. For flies, bees, moths and butterflies the larval stage is when most feeding and growth takes place, so this stage is the most destructive. It is therefore valuable to recognise these pests as larvae and a simple larval key is provided in Table 3.1.

4 Control techniques

4.1 Cultural control

Cultural control is the use of farming techniques or cultural practices associated with crop production, designed to make the environment difficult for the survival of pests. Seldom are such techniques spectacular but they are inexpensive and have no unpleasant side-effects, whilst they often enhance biological control.

4.1.1 Habitat diversification

Whilst tropical subsistence farming may not produce maximum yields it frequently provides relatively pest-free conditions. Small fields surrounded by natural vegetation or hedges are readily accessible to the full compliment of natural enemies. Adjacent vegetation often provides a suitable environment for parts of their life-cycles. Habitat diversification is usually achieved by the planting of more vegetation for use by beneficial organisms. For example, blackberry bushes are planted in the Californian Napa and Sonoma valleys where grapes are grown. The bushes provide a source of overwintering leafhopper eggs of *Diknella cruentata*, for use by the mymarid egg parasite, *Anagrus epos* (Fig 4.1). In summer this parasite attacks eggs of the grape leafhopper, *Erythroneura elegantula*, and can virtually eliminate the third and economically damaging generation of this leafhopper which causes much damage to the leaves and fruit of the vines. The leaves drop in the winter and the grape leafhopper overwinters by adult hibernation at the edge of vineyards. *Anagrus* cannot survive the winter as adults and have no overwintering hosts. Prior to 1961 growers in other valleys, such as Sacremento and San Joaquin, suffered great economic loss until they too began to plant blackberry bushes.

Trials in the USA have shown that crop spraying with a yeast and sucrose mixture helps support populations of natural enemies such as lacewings, hoverflies and coccinellids. A sugar supply is needed to induce the flight behaviour of many parasitoids including *Encarsia formosa*, the parasitoid of the glasshouse whitefly. Normally this is provided by the sweet honeydew secreted by the whitefly themselves, so to start an *Encarsia* colony a few ‘banker’ or infected plants are first introduced to the greenhouse.

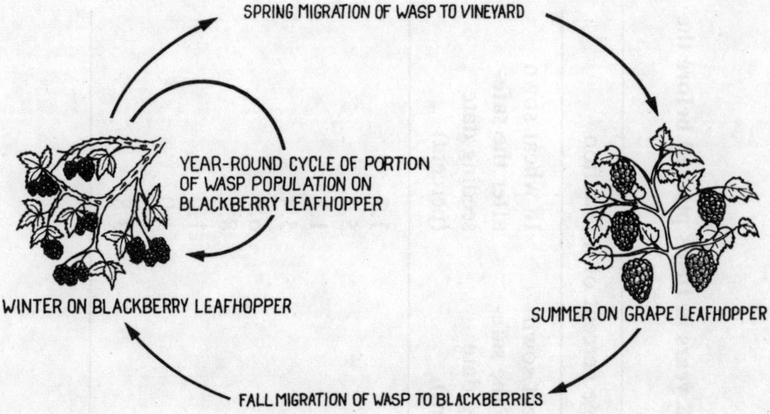

Fig 4.1 Life cycle of *Anagrus epos*, a parasite of the grape leafhopper. The parasite spends its winters in blackberry bushes parasitising eggs of the blackberry leafhopper, while the grape leafhopper is inactive. In the spring, when the grape leafhopper is active again, part of the *Anagrus epos* population migrates back to the vineyard to parasitise the grape leafhoppers' eggs. Some of the parasite population remains in the blackberry bushes throughout the year.

4.1.2 Planting and harvesting dates

Often the planting or harvesting of a crop can be timed to avoid periods of major pest emergence. In winter-wheat growing areas of the Midwest United States, hessian fly adults, *Mayetiola destructor*, which lay eggs for the autumn generation, emerge over a 30-day period during late summer and survive only a few days. Planting is therefore delayed so that germination occurs after all the flies have emerged and died (Table 4.1).

Planting corn early in the northern USA allows maturation before corn earworms and fall armyworms migrate from their southern overwintering sites. In coniferous plantations, the planting of young pine trees is delayed 9 months in the southern USA, and 2–3 years in the north, to avoid damage from the residue of pine root collar weevils of previous stands.

4.1.3 Soil tillage

Ploughing the soil in winter often disturbs overwintering larvae or grubs; they are either crushed, broken-up, eaten by birds, desiccated in the sun or ploughed under too deep for emergence. In North

Table 4.1 Average yields of wheat and percentage of hessian fly infestation over 8 years in fields planted before the safe-seeding date

	Average yield		Average percent of infestation	
Location of field	From wheat sown before the safe-seeding date (bushels/acre)	From wheat sown after the safe-seeding date (bushels/acre)	In wheat sown before the safe-seeding date (percent)	In wheat sown after the safe-seeding date (percent)
Rockford, Illinois	21.8	28.1	24.5	1.7
Bureau, Illinois	27.4	32.9	45.5	5.2
La Harpe, Illinois	30.8	36.5	38.0	1.8
Urbana, Illinois	29.5	37.1	32.6	5.4
Virden, Illinois	23.6	28.4	48.0	6.3
Centralia, Illinois	14.5	21.9	81.0	8.0
Carbondale, Illinois	21.5	23.9	16.0	1.0
Grand Chain, Illinois	15.5	21.4	32.3	1.0
Average	23.1	28.8	39.7	3.8

Dakota, stubble cultivation reduces the spring emergence of wheat-stem sawfly by 75%.

4.1.4 Crop rotation

This technique is most effective against soil inhabiting pests, such as wireworms and white grubs (beetle larvae), which take several years to mature. Altering the available host plants causes large mortality to host-specific pests. For example, legumes in rotation with grass crops reduces both white grub damage to the grasses as well as damage to the legumes by whitefringed beetles (*Graphognathus* spp.). Rotating alfalfa, crimson clover or buckwheat with potatoes reduces wireworm damage to the potatoes.

For many years in the Midwest USA, a 4-year rotation of corn, oats and clover kept corn rootworms, *Crambus caliginosellus*, well below economically damaging levels. The arrival of soil insecticides made this rotation system obselete. Eventually, however, rootworm populations developed resistance to the insecticides and greater doses were applied, resulting in higher soil pesticide residues. Yet, for economic reasons, farmers are still unwilling to return to the 4-year rotation.

4.1.5 Residue disposal

Good sanitation procedures greatly benefit pest control. In South Africa removal of fallen oranges depletes the larval pool of the African false codling moth, *Cryptophlebia leucotreta*, making it easier for the parasitic natural enemy, *Trichogrammatoidea lutea*, to contain the remaining moth population. Removing corn subble is especially important in reducing overwintering stem-boring larvae. In the case of cotton, the removal of old, pest-harbouring bolls is also important. After the cotton harvest in India, goats are allowed to graze in the fields, eating all the remaining infested bolls and stalks.

4.1.6 Irrigation

Although many farmers have little control over the amount of water that reaches the land, in drier areas irrigation paractices can sometimes be regulated to work against pests. On the Pacific slopes of the USA, wireworms may be controlled by either flooding the land for several days, or by allowing it to dry out during the summer months. Flooding vineyards may provide some measure of control against grape Phylloxera, *Viteus vitifoliae*, an aphid-like pest of roots. The pink bollworm on cotton, *Pectinophora gossypiella* (Saunders), is also very sensitive to drought or flood, but flooding often extends the host's life-cycle, providing for an extra generation of the pests.

4.1.7 Trap crops

Bordering a crop with a particularly attractive alternative host may prevent colonisation of the crop. The concentrated pests may then be destroyed by smaller doses of insecticide. In Canada during the 1950s, brome grass was planted in 15–20 m strips around wheat fields. The stem-boring sawfly, *Cephus cinctus*, did not even penetrate to the wheat, leaving all its progeny inside the brome grass. It was not even necessary to destroy the brome grass since many larvae occurred per stem and subsequently they cannibalised each other or were parasitised by the many parasitoids attracted to the high host densities. Very few adult sawflies actually emerged.

4.1.8 Regulatory control

Most countries have strict quarantine laws which are aimed at preventing the introduction of foreign pests.In 1965 the United States Quarantine Service intercepted 32 572 pests of which the major species were: European cherry fruit fly, *Rhagoletis cerasi* (L.), (104); Mediterranean fruit fly, *Ceratitis capitata* (Wiedemann), (196); Mexican fruit fly, *Anastrepha indens* (Loew), (223); Oriental fruit fly, *Dacus dorsalis*, (84); Khapra beetle, *Trogoderma granarium*, (462); potato root eelworm, *Heterodera* sp., (101); sweet organe scab (299) and citrus blackfly, *Aleurocanthus woglumi* (571) [Rainwater, H. I. and Smith, C. A. *Agriculture Yearbook 1966*, US Government printing office, Washington D.C.]. Particularly serious pests may be subject to a 'notification order' whereby any farmer who suspects that the pest has appeared on his crop must notify the appropriate authorities who then carry-out pest eradication procedures. There are also certain cultural practices which are required by law. A sugar beet rotation is in force in Britain to control the beet eelworm, *Loxostege sticticalis*. Cotton planting and stalk destruction dates are enforced in Texas to eradicate the pink bollworm, *Pectinophora gossypiella* (Fig 4.2).

4.2 Physical and mechanical controls

Physical and mechanical methods of pest control differ from cultural techniques in that they are applied directly to the pest. For example, tomato hornworms may be picked directly from tomatoes and flies can be swatted.

4.2.1 Barriers

The simple precaution of screening windows and doors to prevent entry of mosquitoes is well-known to many householders. Similarly,

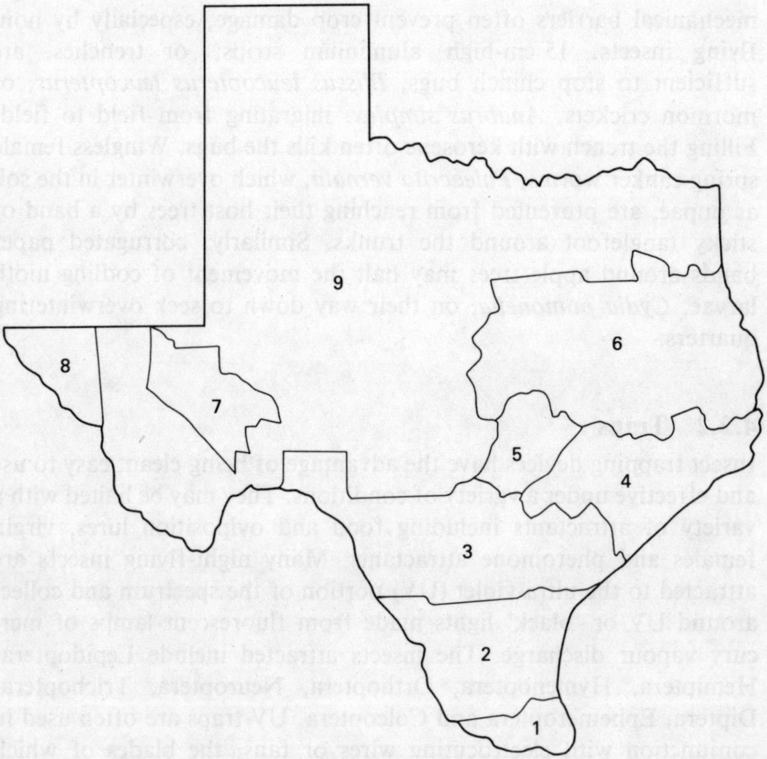

Fig 4.2 Cultural control programme for pink bollworms in Texas showing legally designated cultural control zones, dates of planting, and stalk destruction deadlines for each zone as set by law. Differences in zone planting dates and stalk destruction deadlines are due to climatic differences in each zone affecting pink bollworm emergence in the spring and overwintering diapause in the fall. *Zone 1*: planting period, 1 February–31 March; stalk destruction deadline, 31 August (fall okra cannot be planted in this zone before 15 October). *Zone 2*: planting period, 15 February–20 April; stalk destruction deadline, 25 September. *Zone 3*: planting period, 5 March–10 May; stalk destruction deadline, 20 October. *Zone 4*: planting period, 10 March–20 May; stalk destruction deadline, 20 October. *Zone 5*: planting period, 10 March–20 May; stalk destruction deadline, 31 October. *Zone 6*: planting period, 20 March–31 May; stalk destruction deadline, 30 November. *Zones 7 and 8*: stalk destruction deadline, 1 February. *Zone 9*: no mandatory stalk destruction deadline.

mechanical barriers often prevent crop damage, especially by non-flying insects. 15 cm-high aluminium strips, or trenches, are sufficient to stop chinch bugs, *Blissus leucopterus leucopterus*, or mormon crickets, *Anabrus simplex*, migrating from field to field. Filling the trench with kerosene often kills the bugs. Wingless female spring canker worms, *Paleacrita vernata*, which overwinter in the soil as pupae, are prevented from reaching their host trees by a band of sticky tanglefoot around the trunks. Similarly, corrugated paper bands around apple trees may halt the movement of codling moth larvae, *Cydia pomonella*, on their way down to seek overwintering quarters.

4.2.2 Traps

Insect trapping devices have the advantage of being clean, easy to use and effective under a variety of conditions. They may be baited with a variety of attractants including food and oviposition lures, virgin females and pheromone attractants. Many night-flying insects are attracted to the ultra-violet (UV) portion of the spectrum and collect around UV or 'black' lights made from fluorescent lamps of mercury vapour discharge. The insects attracted include Lepidoptera, Hemiptera, Hymenoptera, Orthoptera, Neuroptera, Trichoptera, Diptera, Ephemeroptera and Coleoptera. UV traps are often used in conjunction with electrocuting wires or fans, the blades of which pulverise the insects. The range over which insects will be attracted to UV varies: it is effective up to about 45 m for pink bollworms, *Pectinophora gossypiella*, and 120 m for tobacco hornworms, *Manduca sexta*. Probably the most successful use of black lights has been in North Carolina where one 15 W trap per acre (0.4 ha) has provided effective control against tobacco hornworms. Other light sources, such as yellow light, are often used as repellents. Aluminium foil beneath gladioli repels winged aphids, the vectors of mosaic virus.

Other successful lures include traps baited with virgin females or pheromones (see section 4.6, page 68). Daytime fliers are often attracted by food baits, often used in conjunction with a poison. Trimedlure (Pherocon MFF®) is a synthetic food bait for the Mediterranean fruit fly, *Ceratitis capitata*, and sprays of yeast, protein and malathion have also been used successfully in Florida against this fly. The oriental fruit fly, *Dacus dorsalis*, is attracted to methyleugenol, a constituent of rotting fruit. Boards impregnated with methyleugenol and an insecticide and dispersed on the island of Rota in the Marianas provided inexpensive and effective control. Ammonia-yielding chemicals mimic decaying organic matter, thereby attracting ovipositing flies. Corn earworm adults, *Heliothus armigera* will oviposit on twine impregnated with corn silk juice.

Despite the variety of attractants, insect traps alone are rarely sufficient to provide adequate control. Usually they are used in conjunction with other methods of management. In Quincy, Florida, black lights mounted on the perimeter of the crop reduced annual insecticide sprays on tobacco from 17 to 2! Perhaps the greatest use of traps, however, is as sampling tools to gauge the effectiveness of other methods of control.

4.2.3 Temperature and humidity control

Temperature and humidity variations are best used against pests of stored products, such as grain, where 3–4 hours at 52–55°C (125–131°F) in a high-frequency electrostatic field kills most pests. On living grain, pest destruction is more difficult since the grain must not also be destroyed.

Low temperatures can also be lethal to pests. One or two days at −22°C (−7°F) kills most insects. Again, where crop preservation is essential, more care must be exercised. 60 days at 0–3°C (32–37°F) kills apple maggots and 33 days at these temperatures is lethal to plum Curculio larvae.

4.3 Host plant or varietal control

It has long been known that some varieties of plants are more resistant to pest attack than others. Crossing resistant with high-yielding varieties can produce a crop that is both pest-resistant and high-yielding. In the late 19th century the root aphid, grape Phylloxera, *Viteus vitifoliae*, was accidentally introduced into Europe. The European grape, *Vitis vinifera*, was so susceptible that the entire wine industry was threatened with ruin. The vineyards were saved in about 1870 by the discovery of pest-resistant American grapes, *V. labrusca*, and by the development of grafting techniques which enabled resistant roots to be joined onto popular European varieties.

Resistance may be provided by one gene (monogenic or vertical resistance) or by many genes (polygenic or horizontal resistance). Monogenic resistance is fairly easy to select for in crosses, and gives good protection; virtual immunity to host-specific pests or strains. The mechanism for selecting resistance through one gene and high yield through another gene is:

High-yielding, suseptible variety × wild strain: resistant but not high-yielding (gene for resistance)

↓

F_1 all resistant since wild gene is dominant

↓ (F_1 selfed)

F_2 Segregation and assortment of gametes according to Mendel's 1st and 2nd laws

↓ (Resistant form backcrossed with original high yielder. Backcrossing continued for 6–7 generations.)

Pure-breeding, high-yielding, resistant variety

Resistant varieties have been developed in this way for wheat, apples, tomatoes, maize and rice.

Polygenic resistance gives less protection against specific pests but, because of its polygenic nature, it generally provides protection against a wider variety of pests, and has less chance of being circumvented by new pest strains. However, since many genes are involved, polygenically resistant forms are much more difficult to select for.

Resistance in plants may be regarded as real or apparent. Apparent resistance includes early-maturing crop varieties which evade peak host densities, or varieties with a low water content. Real resistance varies along a continuum from immunity, to high resistance, to low resistance, to susceptibility, to high susceptibility. Immunity itself is derived from the interaction of three factors (Fig 4.3):

4.3.1 Antibiosis

Antibiosis in plants interferes with the insect's life history, causing reduced life-span, fecundity or size, or increased mortality. It is usually a function of the nitrogen or amino acid concentration of plant tissues. Some plants actually produce ecdysones that act as insect anti-hormones, interfering with juvenile hormone production and so preventing successful metamorphosis. For example the bedding plant, *Ageratum haustoneanum*, contains ecdysones, called Precocene I and II, which induce precocious metamorphosis, shortened life-cycle, reduced feeding and sterile females in the milkweed bug, *Oncopeltus fasciatus*, and cotton stainer, *Dysdercus suturellus*, and induce diapause in the Colorado beetle, *Leptinotarsa decemlineata*.

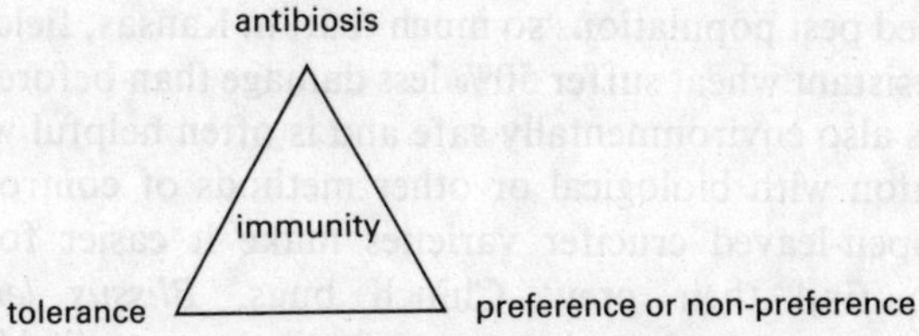

Fig 4.3 The factors that interact to provide immunity

It was recently demonstrated that when glandular hairs of the wild potato, *Solanum berthaultii*, are broken by aphids, the aphid alarm pheromone, (E)–B–farnesene, is actually released by the plant causing the aphids to leave. Intense research is now in progress to incorporate this characteristic in commercial potato varieties.

4.3.2 Tolerance

Tolerance is the ability of the plant either to repair an injury very well or to tolerate the injury.Tolerance is usually shown towards specific pests only. A striking example of injury repair is provided by conifers which exude gums or resins after injury by the pine shoot moth, *Evitria busliana*. Some tea varieties are usually tolerant of damage by the shothole borer, *Xylebotus fornicatus*, and new growth rapidly occurs over the beetle's gallery entrances.

4.3.3 Preference or non-preference

Some plants have features that discourage pests from feeding, ovipositing or sheltering on them. Sticky glandular hairs on potatoes and tomatoes discourage feeding by aphids, which stick to them. Mustard oils in the Cruciferae family generally discourage insect feeding. However, certain cabbage caterpillars use mustard oils as a feeding cue and will only feed on leaves containing, or painted with, these oils.

The most serious problem with varietal resistance is that it takes a long time to develop—between 10 and 15 years—and in the interim new pest strains which circumvent resistance may appear. This is particularly likely with monogenically resistant crop varieties. Circumvention of resistance by pests develops in much the same way as resistance to pesticides. The advantages of varietal resistance are that resistance is usually maintained for long periods and that the effect is cumulative because fecundity is reduced in every successive pest generation. Wheat resistance to the hessian fly, *Mayetiola destructor*, has depressed pest populations so much that, in Kansas, fields planted with non-resistant wheat suffer 50% less damage than before. Varietal resistance is also environmentally safe and is often helpful when used in conjunction with biological or other methods of control. For an example, open-leaved crucifer varieties make it easier for natural enemies to find their prey. Chinch bugs, *Blissus leucopterus leucopterus*, on corn and sorghum are effectively controlled by the use of resistant varieties along with mechanical barriers. Spotted alfalfa aphids, *Therioaphis trifolii*, are controlled by varietal resistance and natural enemies.

4.4 Chemical control

4.4.1 Benefits

Toxic chemicals are still the main defence against pest attack and they are likely to remain so for many years to come.

The appeal of chemical pesticides for use in pest control can be attributed to a number of factors, not the least of which is economy. Many modern insecticides cost only $2–3 US per kg, and have a benefit/cost ratio approaching 4–5. Pesticides can be applied easily, often using a one-man aeroplane. Aeroplanes were used as early as 1921 against the catalpa sphinx moth, *Ceratomia catalpae*, in Ohio. After the second world war, 'planes, pilots and pesticides became readily available so that entire fields could be treated in less than an hour.

Of greatest importance, perhaps, insecticides often afford the only practicable and quick control method where pest populations are approaching economic thresholds. Lethal action is often evident after just a few hours or a day or two. Swarms of migratory locusts, *Schistocerca gregaria* (Forsk.) and *Locusta migratoria migratorioides* (R. and F.), have been aerially sprayed with dinitro-o-cresol (DNOC) and with dieldrin and destroyed in flight, thus protecting entire agricultural regions. Other techniques cannot be used in such emergency situations since they require long-term planning.

Insecticides may be applied in a wide variety of forms: fumigants; smokes; aerosols; sprays; dusts and granules. They may be applied to the soil, used in baits or seed treatments, impregnated into cloth, timbers and paper or administered as systemics to plants. Several insecticides may be used together to achieve a desired range of properties. For an example, a spray of carbaryl, azinphos-methyl and dicofol is used to control apple pests in orchards.

Insecticides remain the single most important control technique for insect vectors of animal and plant diseases (see Chapter 7), particularly plant viruses.

Historically, one of the first examples of chemical control was the successful application of kerosene emulsions to San José scales, *Quadraspidiotus perniciosus*, in California in 1868. In the late 1800s Paris green (an arsenic derivative) was used against Colorado beetles, *Leptinotarsa decemlineata*. Early in the 20th century, apple scab was treated by lime-sulphur, providing blemish-free fruit. In 1918 calcium-arsenic dust was applied to control cotton boll weevils, *Anthonomus grandis*. Cotton has always required a great deal of management. Of the total volume of major insecticides used today, 47% is used on cotton, 17% on corn, 6% on apples, 3% on tobacco and 2% on soybeans. Of the remainder, 25% is used in industry, 15% in the home and garden and 10% by governments.

Without insecticides, it is estimated that half as much land again would be needed to produce the same amount of food as we currently harvest. In the US alone losses of over $2 billion would occur as a result of pest damage if spraying ceased. Cotton leafworm, *Alabama argillacea*, and cotton aphid, *Aphis gossypii*, have been almost eradicated thanks to organophosphate insecticides. Yield increases due to insecticide use are often dramatic. In Ghana, cocoa production has gone up 300%, and cotton production throughout the world has doubled.

4.4.2 Insecticide structure and classification

Pesticides can be referred to by (i) their common names (in the United States these are designated by the Entomological Society of America), (ii) the registered trade name which may differ according to different manufacturers, (iii) the scientific name and (iv) the structual formula (Fig 4.4).

Most modern insecticides work as contact poisons, whereas older arsenic derivatives were stomach poisons. They can be divided into

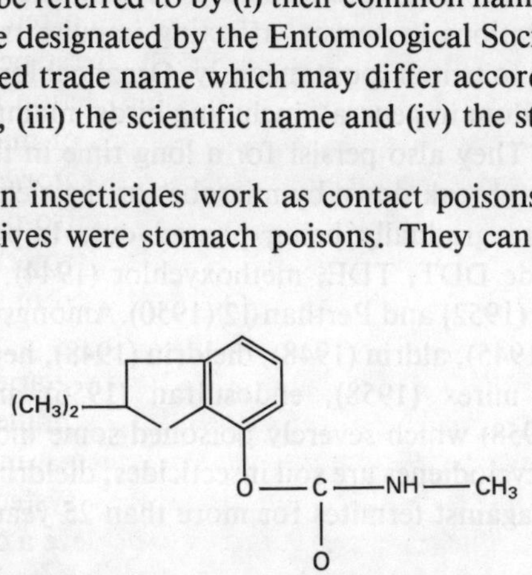

Fig 4.4 This is called (i) carbofuran, (ii) furadan or (iii) 2.3, dihydro-2.2-dimethyl-7-benzafuranyl methylcarbamate.

fumigants, repellents (which are most useful against vectors) and systemics (which operate via the plant sap). Insecticides can be further classified into organics, which constitute 97% of current insecticides, and inorganics (of which the most useful is silica dust which abrades waxes from the cuticle causing dehydration). Organics can be divided into four main categories: the 'hard', persistent organochlorines; the 'soft', less persistent organophospates; carbamates and formamidines. A further group of organics is made up of the 'botanical' insecticides. These are discussed after the four main categories.

Organochlorines

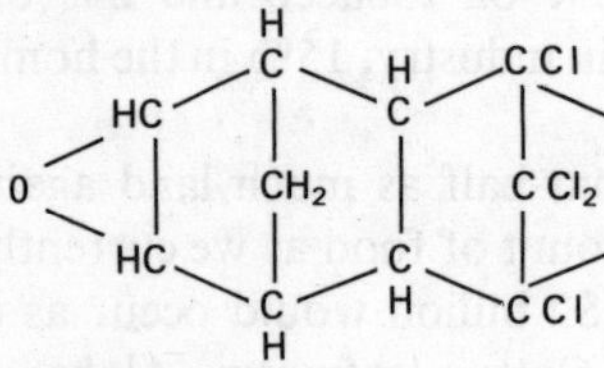

Fig 4.5 Dieldrin

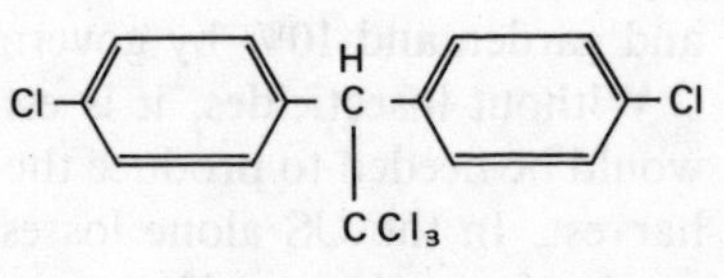

DDT

These compounds contain C, H, O and a Cl group. The most famous organochlorine is probably DDT (dichloro-diphenyl-trichloroethane) which was first made in 1939 and subsequently banned in the US in 1973. Organochlorines are grouped into straight carbon chain forms (diphenyl aliphatics) or ringed structures (cyclodienes). Both alter Na and K concentrations in nerves, affecting impulse transmission and causing muscles to twitch spontaneously. Organochlorines are toxic to the animal kingdom in general, including birds, mammals and fish as well as insects. They also persist for a long time in the environment being difficult to break down by microbes or the elements. For these reasons they are gradually being phased out. Important diphenyl aliphatics include DDT, TDE, methoxychlor (1944), dicofol (1957), chlorobenzilate (1952) and Perthan⑫(1950). Amongst the cyclodienes are chlordane (1945), aldrin (1948), dieldrin (1948), heptachlor (1948), endrin (1951), mirex (1958), endosulfan (1956) and chlordecone (Kepone®) (1958) which severely poisoned some industrial workers in 1976. Many cyclodienes are soil insecticides, dieldrin and chlordane being effective against termites for more than 25 years.

Organophosphates

These take the general formula shown in Fig 4.6 where G_1 and G_2 are often CH_3O or C_2H_5O, A is O or S and Z is one of a wide range of

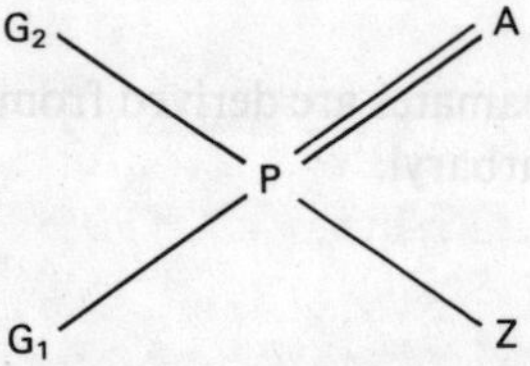

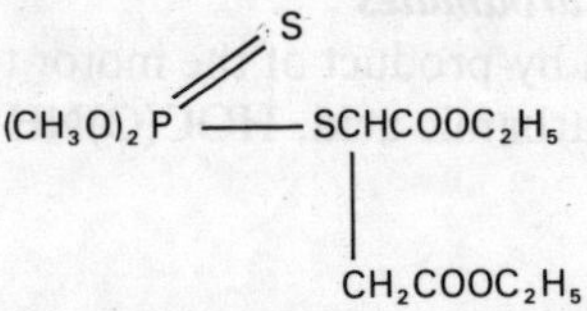

Fig 4.6 General organophosphate Malathion

groups, usually acidic.

Organophosphates are derivatives of phosphoric acid and are related to the 'nerve gases' discovered during the second world war. Like the organochlorines, many organophosphates are equitoxic, but most are less persistent in the environment and so are replacing organochlorines. Organophosphates act as anticholinesterases. They operate by attaching to cholinesterases which normally de-activate acetylcholine (the transmitter of nervous impulses across synapses). This stops the cholinesterases functioning and results in paralysis. The several classes of organophosphates include aliphatic derivatives (usually simple derivatives of phosphoric acid), phenyl derivates (which contain a benzene ring) and heterocyclic derivatives (which contain rings with unlike atoms, e.g. N or O, replacing C).

The many aliphatic derivatives include contact poisons such as malathion (1950), a relatively safe insecticide which may be used indoors, parathion (1949), diazinon (1956), fonofos (1967), trichlorfon (1952) and dichlorvos (Vapona ®) (1960), often an active ingredient in dog and cat flea collars and a good 'knock down' agent. This group also contains many important systemics such as monocrotophos (very toxic to humans) dimethoate (1956), dicrotophos (1963), mevinphos (Phosdrin ®) (1953), oxydemotenmethyl (1960), disulfoton (1956), demeton (Systox ®) (1950) and phorate (1954). Of these, mevinphos is particularly useful since it degrades in less than a day, so crops can be harvested a day after spraying. Systemics are only effective if the insects' intake of plant sap is high, and they are of little use against chewing insects.

Important members of the phenyl derivatives group include ethyl parathion and methyl parathion (1949) which have low toxicity to other animals, the equally safe Gardona ® and the systemics ronnel (Korlan ®) (1954) and crufomate (Ruelene ®) (1959).

Amongst the heterocyclic forms, which are more complex molecules and therefore more persistent, diazinon (1956), and azinphos-methyl (Guthion ®) (1953) are prominent.

Carbamates

A by-product of the motor tyre industry, carbamates are derived from carbamic acid, $HOC(O)NH_2$, for example, carbaryl:

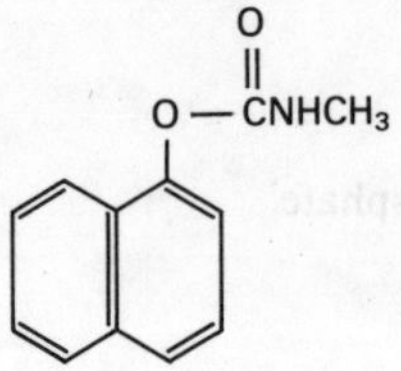

Fig 4.7 Carbaryl

Their mode of action is similar to the organophospates in that they inhibit cholinesterase, but they have lower oral and dermal toxicity and hence are more popular. Carbamates are often used as plant systemics, for example methomyl (1967), aldicarb (Temik ®) (1965) and carbofuran (Furadan ®) (1969).

Despite the fact that carbamates are generally poor acaricides they include many useful insecticides such as carbaryl (Sevin ®) (1956), mexacarbate (Sectran ®), formetanate (Carzol ®) and propoxur (Baygon ®) (1959).

Formamidines

Hailed as the new class of insecticides for the future, formamidines are effective for organophosphate- and carbamate-resident pests. They inhibit monoamine oxidase, allowing biogenic amines to accumulate, which then act as chemical transmitters at synapses, causing continuous nervous transmission. Valued as ovicides and larvicides, important members of this group include chlordimeform (Galecron ® or Fundal ®) (1969) and U-36059. Progress in the development of this group has been hampered by suspected carcinogenic properties.

Botanical insecticides

Many plants contain natural insect toxins, and the so-called botanical insecticides are derived from them, or are their synthesised analogs. Their advantages are that they break down rapidly and are generally environmentally safe.

Pyrethrins are extracted from the flower heads of two types of chrysanthemums. The varieties grown in the Kenyan highlands yield the highest percentage of active ingredients. Pyrethrins cause insect paralysis, and, due to their low mammalian toxicity, are used as 'knock-down' agents in aerosols. The method of production involves much manual labour, so the commercial product is expensive. However, the introduction of synergists, such as piperonyl butoxide, has allowed the content of active pyrethrins to be much reduced without loss of insecticidal activity. More importantly, pyrethrins are rapidly oxidised in sunlight and are thus of limited use in crop husbandry. Recently-produced synthetic analogs, called pyrethroids, offer a solution to both problems, being less labour-intensive and more photostable. The first synthetic pyrethroid, allethrin (1949) is still much used particularly in mosquito coils. Resmethrin and bioresmethrin (1967) are even more active aganist insects than the natural compounds, and have even lower toxicity to mammals. Permethrin (1973) and cypermethrin (1974) are much more photostable than pyrethrum and as such find use as field pesticides. Cypermethrin and deltamethrin (1974) are particularly potent pyrethroids.

Rotenone is derived form tropical legume roots of the genera *Derris, Lonchocarpus* and *Tephrosia*. First used as a fish poison by South American Indians, small doses of rotenone are not toxic to homoiothermic (warm-blooded) animals.

Nicotine, a toxic alkaloid from tobacco, acts as an acetylcholine blocker but is also toxic to homoiotherms. Despite this, it is often sprayed as nicotine sulphate, containing 40% of the alkaloid. It is used as a fumigant in greenhouses.

4.4.3 Improving selectively

It is clearly desirable that as much insecticide as possible reaches the pest population, while as little as possible reaches other non-target organisms or pollutes the enironment (see next section). A selective method of control that is only 50% effective may actually prove better than a non-selective method which kills 90% of the pests and 90% of its natural enemies, although this depends on the degree of mortality exerted by the natural enemies. Selectivity is influenced by insecticide formulations and application methods, insecticide type, insecticide concentration and by the behaviour of the insect itself.

Insecticide formulations and application methods

The active ingredients of insecticides are usually incorporated into stable formulations which are safe to store and most easy to use. These formulations are best stored in containers exhibiting labels indicating (i) the directions for use, (ii) the target insects and crops, (iii) the re-entry statements (usually 1–2 days), (iv) treatments for human contamination and (v) whether the formulation may be applied by the general public or only by registered applicators (Fig 4.8).

The most common formulations include fumigants, granulars, dusts and sprays (Table 4.2). Fumigants have good penetrating ability and types such as methyl bromide can kill eggs and larvae inside plant tissue, as well as adults. Granulars in the soil provide good, slow release of systemic poisons. By far the most important formulations are sprays, with over 75% of insecticides being applied in spray form —an emulsion of oil droplets in water.

Aerial spraying can cover a wide area but is sometimes impractical owing to high winds which cause drift, and the presence of obstacles such as powerlines or high vegetation. Ground application gives a more thorough coverage, but is more labour intensive and it is often difficult to produce fine enough spray droplets. Ultra low volume sprays are often applied as tiny droplets which are either forced out

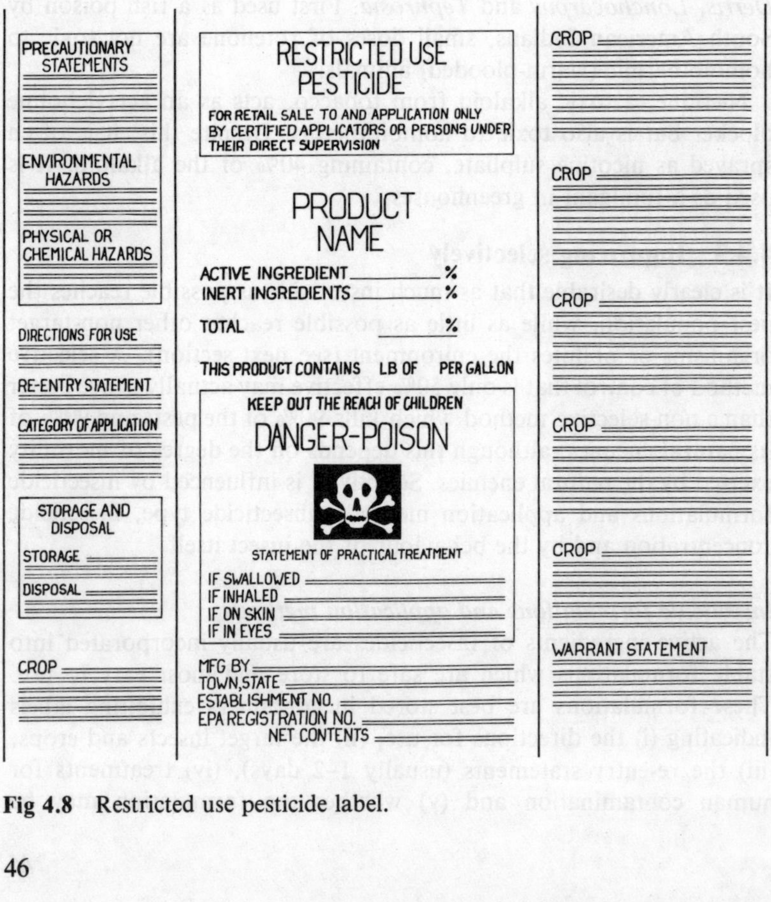

Fig 4.8 Restricted use pesticide label.

Table 4.2 Common formulations of insecticides

1 Sprays
  a Emulsifiable concentrates (EC)
  b Water-miscible liquids (S)
  c Wettable powders (WP)
  d Flowable suspensions (F)
  e Water-soluble powders (SP)
  f Ultra-low volume concentrates (ULV)
2 Dusts (D)
  a Undiluted toxic agents
  b Toxic agents with active dilutent, e.g. sulphur
  c Toxic agents with inert dilutent, e.g. pyrophyllite
3 Granulars (G)
4 Soil fumigants
5 Baits (B)
6 Animal systemics
7 Fertiliser-insecticide combinations (FM)
8 Encapsulated insecticides

under pressure, or projected by centrifugal force from a spinning disc. However, these sprays may also drift from the target. A new technique is now being developed to charge sprays electrostatically. A small droplet with a high charge sticks to plant apices, whilst larger droplets with low charges drift to the ground.

Despite the drawbacks mentioned above, 50–80% of sprays still manage to reach their target area. Other formulations are much more prone to drifting and are less amenable to aerial application. Dusts reach only about 10–40% of their target.

It is vital to achieve good target coverage because, even with 70% or so of pesticide on target, less than 1% of the pesticide application actually filters down to the insects (Fig 4.9). Spraying technique can also be of great importance. On cotton, bollworm pests feed on the lower two-thirds of the plant where the bolls are located, whilst their eggs are laid on plant apices. Sprayers commonly have their 'over-the-row' nozzle blocked leaving only the two side nozzles for spraying. This results in spray reaching the lower two-thirds of the plant only, which permits the build-up in the apices of pirate bug populations which eat the bollworm eggs.

Insecticide type

Although many insecticides are equitoxic, the unique biochemistry

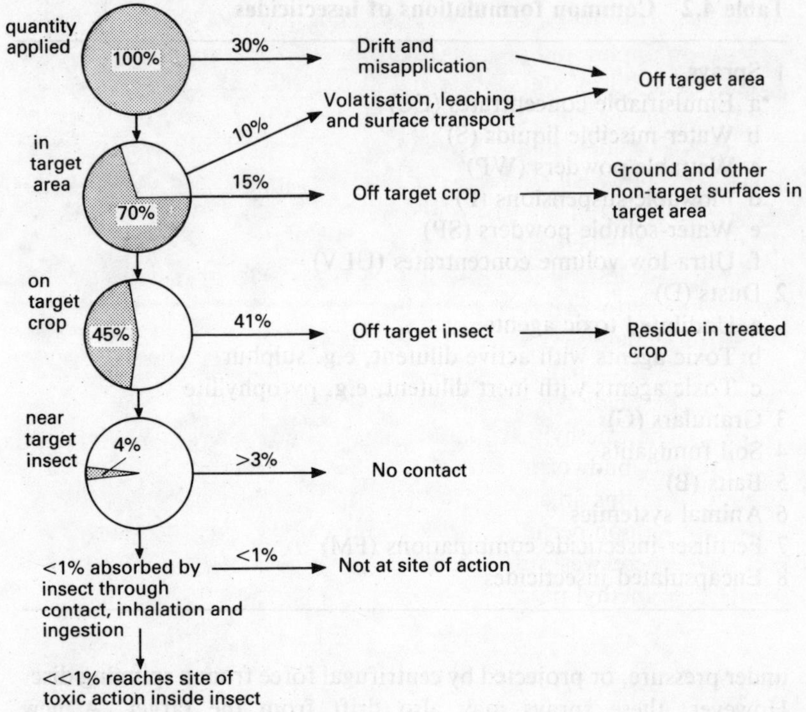

Fig 4.9 Aerial foliar insecticide application: typical losses between spray nozzle and site of toxic action.

of insects causes different species and varieties to vary in their susceptibility to some insecticides. Some organophosphates, for example malathion, contain a P═S group, which in insects is rapidly converted to a P═O group which is the toxin that attaches to cholinesterase. Although the P═O group is also lethal to mammals, the P═S→P═O conversion proceeds very slowly. In addition, mammals can de-esterify the P═S group to an acid, which is quickly eliminated. Insects can carry out his process but only very slowly.

Diflubenzuron affects the synthesis of chitin, a component of insect exoskeletons. It is lethal to lepidopteran larvae that ingest it because it prevents the final moult leading to adulthood. Unfortunately there are few specific insecticides, although systemics are, of course, selective for sap-feeders.

Insecticide concentration

The complexities of choosing the correct application concentrations are well illustrated by attempts at codling moth, *Cydia pomonella*, control on apples and pears in Oregon. Azinphos-methyl at 1.7–2.2 kg/ha was used to control this moth up until 1964. However, it also destroyed the predaceous mites which feed upon a second pest, the two-spotted spider mite. The spider mite problem became so severe that in one test field-spraying was stopped. Predaceous mite populations increased and the pest mite populations peaked annually at 17, 2.5, 2.5 and 0.18 per leaf over the following four years. Codling moths proceeded to destroy 40% of the fruit. Insecticide was re-introduced at 0.5 kg/ha to control the moths, but still the predaceous mites perished. Finally, concentrations were lowered to 0.14 kg/ha, permitting the predatory mites to survive and providing barely adequate pesticide coverage against the codling moth.

In Arizona, methyl parathion at 1.1 kg/ha stops cotton bollworm and tobacco budworm. Unfortunately the most popular marketed insecticide in this area is a mixture of 0.5 kg ethyl parathion/0.28 kg methyl parathion per litre. When 1.4 l insecticide mixture per hectare was used there was sufficient toxicant to kill natural enemies but insufficient methyl parathion (0.397 kg/ha) to kill the pests.

Insect behaviour

Insecticide selectivity can be substantially enhanced by taking insect behaviour into account when applying insecticides. The honey bee, *Apis mellifera*, is very susceptible to insecticides such as methyl parathion, azinphos-methyl and carbaryl. It is best, therefore, to spray fruit trees after petals fall, or at least in the evening when bees are not foraging.

The gravid melon fly, *Dacus euccurbitae*, enters tomato fields to oviposit, but departs at dusk to spend the evening on adjacent vegetation. Application of parathion to such vegetation reduced infestation of fruits from 65% to 3% and reduced hazards to pickers.

Female *Anopheles* mosquitoes prefer to digest their meals of blood while resting in the dark corners formed between walls and ceilings, so they are highly susceptible to spot treatments of DDT.

Perhaps the most astonishing example of selectivity is provided by control of the tsetse fly, *Glossina swynnertoni*. This insect was eradicated over a 90 km^2 area of Africa by selective treatment of resting places with 3% edosulfan or dieldrin. The insecticide was applied to the underside of tree branches that were 2–10 cm in diameter, 1–3 m above the ground and inclined to less than 35° from the horizontal.

4.4.4 Consequences

It is not generally recognised that many insecticides, with rat oral LD_{50} values of < 15 mg/kg, are among the most toxic chemicals known. These include aldicarb, azinphos, carbofuran, demeton, disulfoton, endrin, fensulfothion, mevinphos, parathion, phorate and tetraethyl pyrophosphate.

A pest management rating has been devised for the common insecticides used on agricultural crops in the US in regard to their safety and overall effects on the environment (Table 4.3). Ratings

Table 4.3 Pest management rating of widely-used insecticides

Insecticides	Mammalian Toxicity	Non-target toxicity Fish	Pheasant	Bee	Average	Environ-mental Persistence	Overall Rating
Aldicarb	5	3	5	5	4.3	3	12.3
Aldrin	4	4	4	4	4.0	5	13.0
Azinphos-methyl	4	3	2	4	3.0	3	10.0
Carbaryl	2	1	1	4	2.0	2	6.0
Carbofuran	5	2	5	5	4.0	3	12.0
Carbophenothion	4	2	4	4	3.3	2	9.3
Chlordane	2	3	2	2	2.3	3	7.3
DDT	3	4	2	2	2.7	5	10.7
Demeton	5	2	5	2	3.0	2	10.0
Diazinon	3	2	5	4	3.7	3	9.7
Dicofol	2	1	2	1	1.3	4	7.3
Dieldrin	4	4	3	4	3.7	5	12.7
Dimethoate	3	1	4	5	3.3	2	8.3
Disulfoton	5	3	5	2	3.3	3	11.3
Chlorpyrifos	3	3	3	5	3.7	3	9.7
Endosulfan	4	4	2	2	2.7	3	9.7
Endrin	5	5	5	2	4.0	5	14.0
Ethion	3	2	3			2	7.0
EPN	4	2	3	4	3.0	4	11.0
Gardona	1	4	1	4	3.0	1	5.0
Heptachlor	4	3	4	4	3.7	5	12.7
Lindane	3	3	2	4	3.0	4	10.0
Malathion	2	2	1	4	2.3	1	5.3
Methoxychlor	1	3	1	1	2.3	2	5.3
Methyl parathion	4	1	5	5	3.7	2	9.7
Mevinphos	5	3	5	4	4.0	1	10.0
Naled	2	2	3	4	3.0	1	6.0
Parathion	5	2	4	4	4.0	2	11.0
Phorate	5	4	5	2	3.7	3	11.7
Phosphamidon	4	1	5	3	3.0	2	9.0
Tetraethyl pyrophosphate	5	2	5	3	4.0	1	10.0
Toxaphene	3	4	4	4	3.0	4	10.0
Trichlorfon	2	1	2	1	1.3	1	4.3
Zectran	4	1	5	5	3.7	2	9.7

For explanation of ratings, see text.

were made according to LD_{50} toxicity values for a variety of organisms such as rats, pheasants, rainbow trout and honeybee, and according to their environmental persistence. Environmental persistence ratings are 1 = 1 month, 2 = 1–4 months, 3 = 4–12 months, 4 = 1–3 years and 5 = 3–10 years. Toxicity and environmental ratings were combined to give overall ratings. Following such analysis it was recommended that insecticides with *overall* ratings of between 4 and 7 were suitable for general use. Those rated between 8 and 10 could be used under skillful supervision, whilst those rated between 11 and 13 could be used only under restricted conditions. Insecticides rated above 13 (aldrin, dieldrin, endrin, heptachlor) should find little use in pest management.

Affects on health

In the United States it is estimated there are about 200 deaths annually from insecticide poisoning, and it can be assumed that for each death there are probably 100 non-fatal poisonings. Many developed countries such as the USA, Great Britain, Canada and Sweden have banned or severely restricted the use of persistent insecticides, including DDT. In developing countries, these pesticides still offer the cheapest and most effective means of control. It is now estimated that the total amount of insecticide on the earths surface may reach 1 000 000 tonnes. In 1963, 300 residents of San Joaquin, Bolivia, died of haemorrhagic fever, or black typhus, in an epidemic which was unprecedented. The reservoir of the virus was found to be a mouse-like rodent called a Laucha, large numbers of which had recently appeared in the town. Their arrival coincided with a sudden drop in the cat population, from several 100 to less than 12 in five years. As part of a malaria eradication programme, house interiors had been sprayed with DDT. Cats had picked-up DDT on their fur and had ingested it during grooming, eventually acquiring lethal doses. With these predators removed the Laucha population grew, and with it the virus reservoir.

Persistence and biomagnification

Although mammals are not very susceptible to poisoning by dermal contamination from pesticides, ingestion can cause fatal results. The slow break-down and long persistence of many early insecticides, particularly DDT, led to the build-up in concentration along the food chain, eventually resulting in lethal doses for top predators (Table 4.4).

Table 4.4 Concentrations of organochlorine insecticides in Lake Michigan (1966)

Substrate	Residues (ppm)
Bottom sediment	0.0085
Small invertebrates	0.41
Fishes	3.0–8.0
Herring gulls	3177

Resistance

Even the largest insecticide doses fail to kill some members of a pest population. A very few resistant individuals survive and flourish. Insecticide applications have a dramatic initial effect, killing most individuals. After a few generations, a whole new resistant population arises from the few initially resistant parents (Fig 4.10). Resistance results largely from selection of pre-adapted mutants, possessing genetically controlled mechanisms for detoxfication or target site insensitivity. The resistant housefly, *Musca domestica*, possesses DDT-ase which converts DDT to the inactive DDE. These genetic factors may be present in very low frequencies in the population

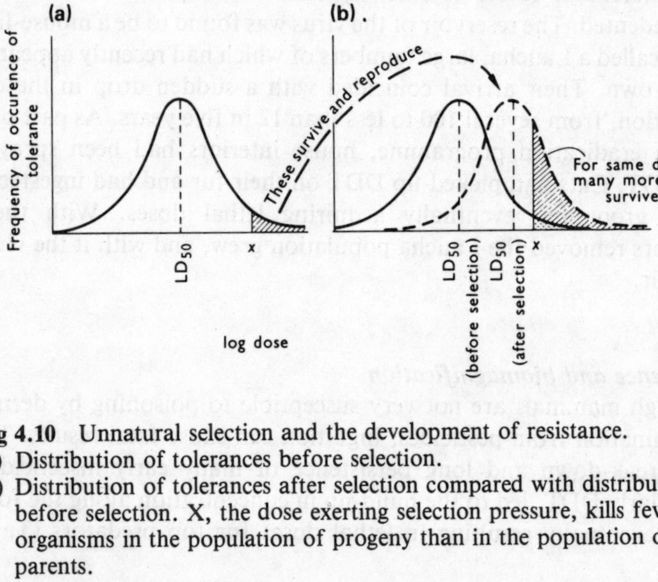

Fig 4.10 Unnatural selection and the development of resistance.
(a) Distribution of tolerances before selection.
(b) Distribution of tolerances after selection compared with distribution before selection. X, the dose exerting selection pressure, kills fewer organisms in the population of progeny than in the population of parents.

before insecticidal treatment. For example, 0.4–6.0% of the *Anopheles gambidae* population in northern Nigeria was found to contain heterozygotes for the single gene of dieldrin resistance. Intensive selection, by dieldrin or BHC residual house spraying, increased the frequency of the gene for resistance to 90% after 1–3 years.

Insects that have developed resistance to one organochlorine insecticide are frequently resistant to another, to which they have not necessarily been previously exposed. This cross resistance is due to the chemical similarity and mode of action of many pesticides.

In 1944 only 44 insect species were known to be resistant to some insecticides; today the figure is over 250. One of the classic cases of insect resistance involves cotton pests in Peru's Canete Valley. This is a naturally arid valley that has been made 'rich' by man-made irrigation. During most of the 1940s, when only arsenicals and nicotine sulphate were in use, the yield was 517 kg/ha. In 1949, however, there were severe outbreaks of cotton bollworms and aphids and yields dropped to 362 kg/ha. The decision to use the new pesticides, DDT, BHC and toxaphene, seemed well-founded, since between 1949 and 1954 the cotton output doubled to 705 kg/ha. However, during this period, the natural enemies of cotton pests were decimated, and resistance to these new chlorinated insecticides was soon developed by the pests themselves. By 1956 the bollworms and aphids had won and the yield dropped to zero. From then on, a more integrated approach was adopted involving less frequent spraying and more reliance on natural enemies; yields climbed to more than 770 kg/ha.

Effects on natural enemies

Insecticides usually have a more severe effect on natural enemies than on the pests themselves, and 'resistant' strains are much less common among natural enemies. Natural enemies are generally free-ranging and come into more frequent contact with insecticide residues than pests, which often shelter on the under-sides of leaves. Furthermore, any surviving predators are left little prey to feed on after insecticide applications, and may starve to death. A resistant strain of natural enemy is, therefore, very difficult to establish.

An important consequence of natural enemy elimination is the development of 'secondary' pests which, in the absence of natural control, are able to reach economically injurious levels. There are numerous examples of this in the literature. In California, an obscure insect, the cotton leaf perforator, *Bucculatrix thurberiella*, became a severe pest following applications of carbaryl to eradicate the pink bollworm, *Pectinophora gossypiella*. The carbaryl virtually eliminated the leaf perforators natural enemies.

The redbanded leaf roller, *Argyrotaenia velutinana*, was a casual or rare pest of apples in the eastern United States and Canada, until spray programmes using DDT, parathion, TDE and malathion disrupted its complex of natural enemies and the leaf roller became a major apple pest.

The vedalia beetle had controlled cottony cushion scale on Californian citrus virtually single-handedly since its introduction in 1888. In 1946 and 1947, many thousands of acres developed damaging populations of scale after DDT application for other pests.

The answer is to develop a more selective insecticide. However, pesticides are extremely expensive to develop. Most companies prefer to market 'broad-spectrum' sprays, from which they stand to gain maximum returns (Fig 4.11). For six years money is expended in pesticide development and another six years on the market is then required just to break even. This leaves only three years for profits before the patent ends.

Contamination of the environment

The 'spreadability' of insecticides from their site of application is a major cause for concern. Insecticides such as DDT, DDE, chlordane

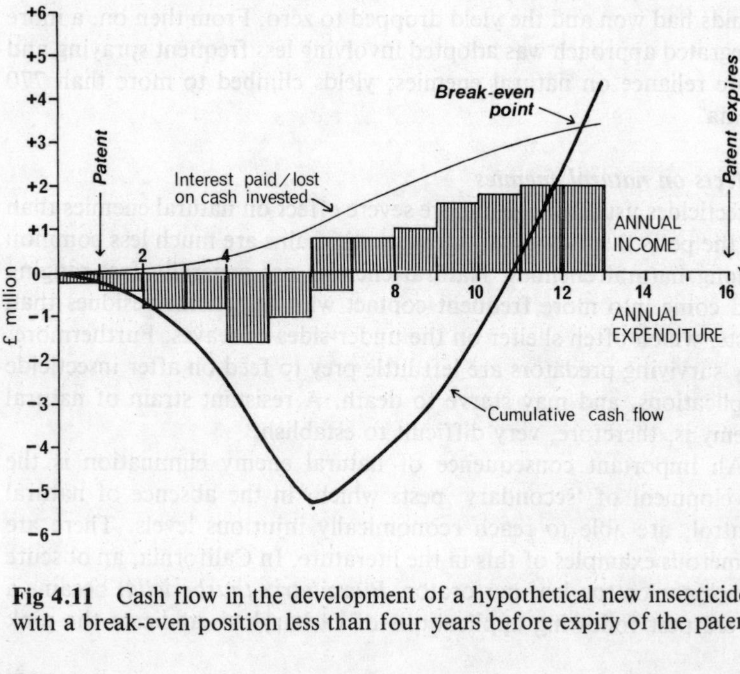

Fig 4.11 Cash flow in the development of a hypothetical new insecticide, with a break-even position less than four years before expiry of the patent.

and dieldrin, that were sprayed in western Texas were picked up by a large dust-storm and later deposited 1500 miles away in Cincinnati, Ohio (Fig 4.12).

Significant quantities of insecticides have been detected in Antarctic penguins inhabiting areas where no insecticide has been applied within several thousand miles.

Such environmental contamination can often be disastrous. In 1963 the Louisiana commercial fishing industry nearly shut down because of fish poisoning caused by run-off of insecticide into the Mississippi. In 1976 the commerical and sports fishing industries of Chesapeake Bay, off the Virginia coast, were shut down owing to pollution caused by the manufacture of the insecticide Kepone ®. Also in 1976, mirex contaminated lake Ontario making the fish unfit for consumption.

Miscellaneous side-effects

The side-effects of spraying vast areas with some of the most deadly poisons known are many, varied and probably still not fully appreciated. Apart from general effects on human health and on other

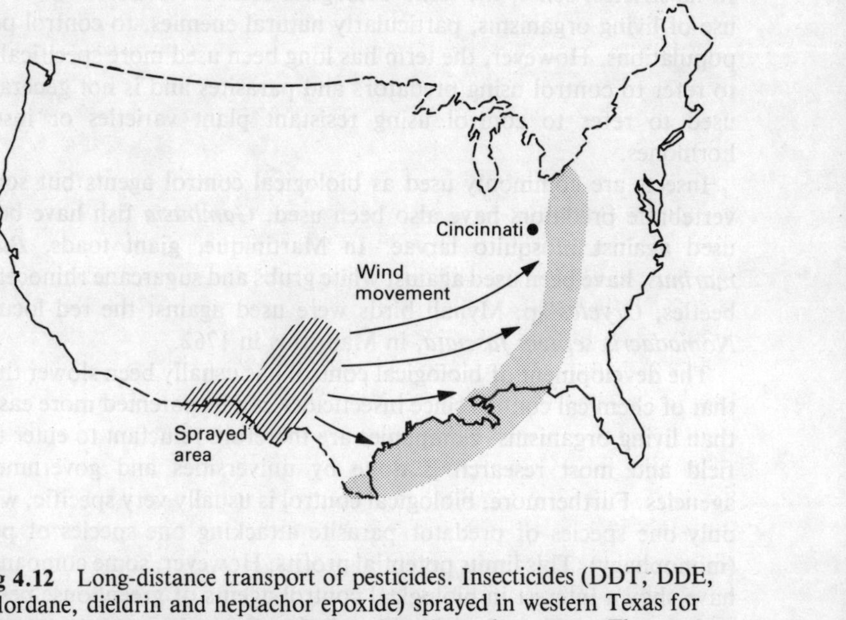

Fig 4.12 Long-distance transport of pesticides. Insecticides (DDT, DDE, chlordane, dieldrin and heptachor epoxide) sprayed in western Texas for agricultural pest control were picked up by a large dust storm. They were deposited in Cincinnati, Ohio, with a little rain the following day at a level of 1.3 ppm. At the time the dust fall occurred in Cincinnati, the dust cloud resulting from the storm stretched nearly 2500 km and was 320 km wide.

non-target organisms, insecticide use may encourage increased densities of bacteria and fungi that feed on pesticide residues in the field, robbing them of their potency. In particular, slow-release granules provide microbes with a source of food for weeks.

Some insecticides may actually harm the crops they are designed to protect. Sulphur-shy currant bushes drop their leaves when a sulphur-based pesticide is used. Usually such properties are detected during field trials, and the chemical is often transferred to the herbicide division!

Finally, DDT blocks calcium metabolism in birds resulting in decreased shell thickness and calcium content of eggs (Fig 4.13). Increases in shell breakages resulted in a decline in the numbers of offspring of some top predators during the 1940s. Those affected included the peregrine falcon, the sparrow-hawk, the brown pelican, the golden eagle, the bald eagle and the osprey.

4.5 Biological control

In its strictest sense, the term 'biological control' refers to the direct use of living organisms, particularly natural enemies, to control pest populations. However, the term has long been used more specifically, to refer to control using predators and parasites and is not generally used to refer to control using resistant plant varieties or insect hormones.

Insects are commonly used as biological control agents but some vertebrate predators have also been used. *Gambusia* fish have been used against mosquito larvae. In Martinique, giant toads, *Bufo marinus*, have been used against white grubs and sugarcane rhinoceros beetles, *Oryctes* sp. Mynah birds were used against the red locust, *Nomadacris septem fasciata*, in Mauritius in 1762.

The development of biological control has usually been slower than that of chemical control since insecticides can be patented more easily than living organisms. Companies are therefore reluctant to enter the field and most research is done by universities and government agencies. Furthermore, biological control is usually very specific, with only one species of predator/parasite attacking one species of pest (monophagy). This limits potential profits. However, some companies have shown interest in biological control agents of greenhouse pests, and parasites such as the whitefly parasite, *Encarsia formosa*, are now on the market.

The use of biological control can be extremely profitable. In New Zealand, in 1974, the importation of a parasitic wasp to control the

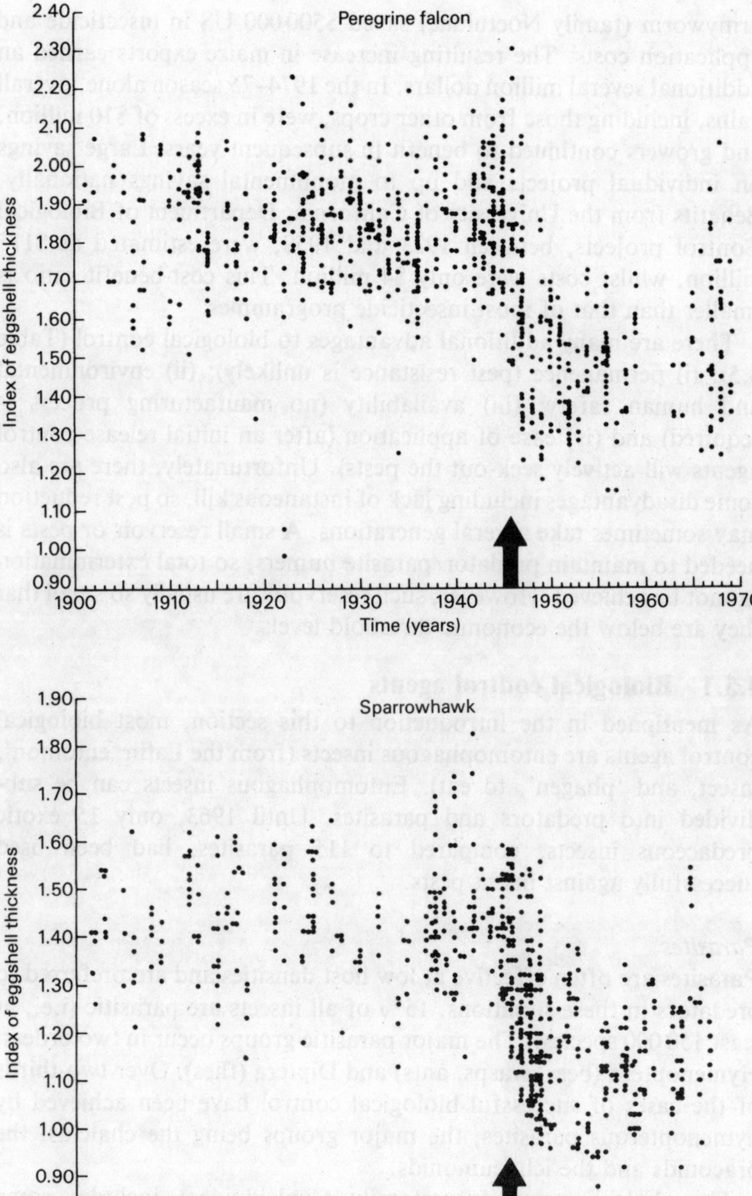

Fig 4.13 Changes in thickness of eggshells of peregrine falcon and sparrowhawk in Britain. Arrows indicate first widespread use of DDT.

armyworm (family Noctuidae) saved $500 000 US in insecticide and application costs. The resulting increase in maize exports earned an additional several million dollars. In the 1974–75 season alone, overall gains, including those from other crops, were in excess of $10 million, and growers continued to benefit in subsequent years. Large savings on individual projects add up to monumental savings nationally. Benefits from the University of California, Department of Biological Control projects, between 1923 and 1959, were estimated at $115 million, whilst costs were only $4 million. This cost-benefit ratio is smaller than that of most insecticide programmes.

There are many additional advantages to biological control (Table 4.5): (i) permanence (pest resistance is unlikely); (ii) environmental and human safety; (iii) availability (no maufacturing process is required) and (iv) ease of application (after an initial release control agents will actively seek-out the pests). Unfortunately, there are also some disadvantages including lack of instaneous kill, so pest reduction may sometimes take several generations. A small reservoir or pests is needed to maintain predator/parasite numers, so total extermination cannot be achieved. However, such reservoirs are usually so small that they are below the economic threshold levels.

4.5.1 Biological control agents

As mentioned in the introduction to this section, most biological control agents are entomophagous insects (from the Latin 'entomon', insect, and 'phagen', to eat). Entomophagous insects can be subdivided into predators and parasites. Until 1963, only 15 exotic predaceous insects, compared to 115 parasites, had been used successfully against insect pests.

Parasites

Parasites are often effective at low host densities and are preferred to predators in these situations. 15% of all insects are parasitic (i.e., at least 150 000 species). The major parasitic groups occur in two orders: Hymenoptera (bees, wasps, ants) and Diptera (flies). Over two-thirds of the cases of successful biological control have been achieved by hymenopterous parasites, the major groups being the chalcids, the braconids and the ichneumonids.

The chalcid group (superfamily Chalcidoidea) includes some twenty-odd families and thousands of species. Most are tiny insects, less than 2 mm long, which develop primarily in host eggs, so having an important regulatory role on host populations (Fig 4.15). Chalcids attack species in nearly all orders but particularly those in the orders Coleoptera, Diptera, Homoptera and Lepidoptera, which also include

Table 4.5 A comparison of the advantages and disadvantages of biological and chemical pest control

Category	Biological control	Chemical control
Environmental pollution: danger to man, wildlife, other non-target organisms, soil, etc.	None	Considerable
Upsets in natural balance and other ecological disruptions	None	Common
Permanency of control	Permanent	Temporary – must apply at least once a year.
Development of resistance to the mortality factor	Extremely rare	Common
General applicability to broad-spectrum pest control	Theoretically unlimited but not expected to apply in practice to all pests. Still underdeveloped.	Can be applied, empirically, to nearly all insects but unsatisfactory with some.
Immediacy of control	Initial control may take 1–2 years, but pest populations remain reduced.	Can rapidly reduce outbreaks, but they recur. Psychologically satisfying to the user at first.

most of our chief crop pests. The families Encyrtidae and Aphelinidae have been particularly outstanding control agents. They have been used against scale insects, mealybugs and whiteflies.

The braconids and ichneumonids both belong to the superfamily Ichneumonidae. The preferred hosts of braconids are lepidopterous larvae, coleoptrous larvae, some dipteran larvae and some homopterans (in particular aphids, whose parasitised appearance has been termed 'mummy'). Parasitism may be external or internal, solitary or gregarious. The type of parasitism is correlated to host habits;

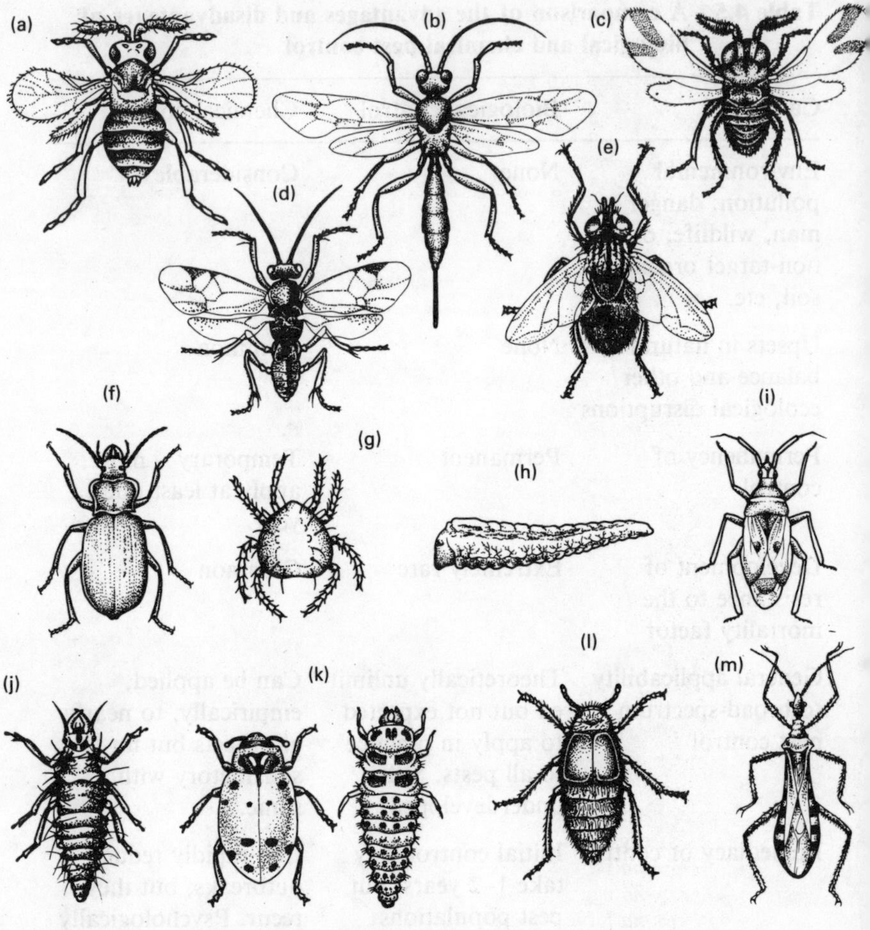

Fig 4.14 Some biological control agents:
(a) an insect egg parasite, *Trichogramma* sp.;
(b) an ichneumon wasp, *Horogenes punctorius*, a parasite of the European corn borer;
(c) a larval parasite of some scale insects, *Comperiella bifasciata*;
(d) a braconid, *Apanteles* sp., a parasite of Lepidopteran larvae;
(e) a tachinid fly, *Winthemia quadripustulata*, an armyworm parasite;
(f) a ground beetle, *Calosoma sycophanta*, a parasite of Lepidopterans;
(g) a predatory mite, *Phytoseiulus persimilis*;
(h) a hover fly larva, *Syrphus* sp., an aphid predator;
(i) an anthocorid bug, *Anthocoris nemorum*, a mite predator;
(j) a lacewing larva, *Chrysopa carnea*, a general insect predator;
(k) adult and larva of a ladybird beetle, *Rodolia cardinalis* (the vedalia beetle), a citrus cottony cushion scale predator;
(l) a staphylinid beetle, *Oligota* sp., a spider mite predator;
(m) an assassin bug, family Reduviidae, a general insect predator.

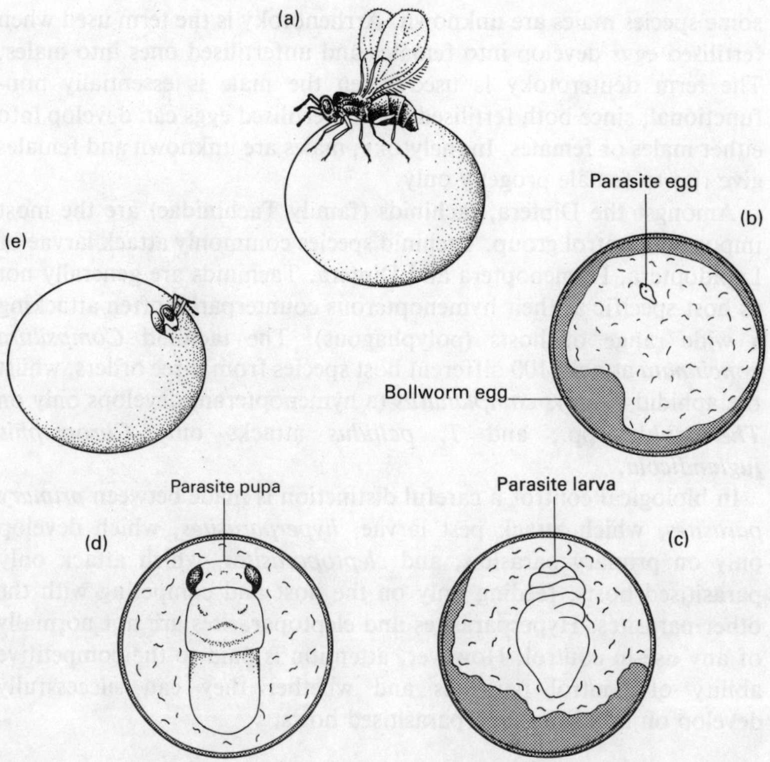

Fig 4.15 Stages in the life-history of *Trichogramma* sp., a solitary endoparasite of eggs.
(a) *Trichogramma* female ovipositing in an insect host egg.
(b) *Trichogramma* egg within the host egg.
(c) *Trichogramma* larva within the host egg.
(d) *Trichogramma* pupa within the host egg.
(e) adult *Trichogramma* emerging through hole it has cut in the host eggshell.

external parasitism tends to occur in hosts living in protected places, such as tunnels, and internal parasitism is more common in free-living hosts. There are more species of ichneumonid than there are species of vertebrate, the Ichneumonidae composing about 20% of all parasitic insects. Most of these probably parasitise wood- and stem-boring larvae in the Lepidoptera, Coleoptera or Hymenoptera orders, but some have very diverse habits.

A particularly appealing feature of many Hymenoptera is their ability to maintain populations in the absence of males. Indeed, in some species males are unknown. Arrhenotoky is the term used when fertilised eggs develop into females and unfertilised ones into males. The term deuterotoky is used when the male is essentially non-functional, since both fertilised and unfertilised eggs can develop into either males or females. In thelytoky, males are unknown and females give rise to female progeny only.

Amongst the Diptera, tachinids (family Tachinidae) are the most important control group. Tachinid species commonly attack larvae of Lepidoptera, Hymenoptera and Diptera. Tachinids are generally not as host-specific as their hymenopterous counterparts, often attacking a wide range of hosts (polyphagous). The tachinid *Compsilura concinnata* attacks 100 different host species from three orders, whilst the aphidid *Trioxys complanatus* (a hymenopteran) develops only on *Therioaphis* spp., and *T. pallidus* attacks only *Chromaphis juglandicola*.

In biological control, a careful distinction is made between *primary parasites*, which attack pest larvae, *hyperparasites*, which develop only on primary parasites, and *cleptoparasites*, which attack only parasitised hosts, feeding only on the host and competing with the other parasites. Hyperparasites and cleptoparasites are not normally of any use in control. However, attention is paid to the competitive ability of control parasites and whether they can successfully develop on multi- or superparasitised hosts.

Predators

Predators are especially useful at high host densities, individual predators destroying (by eating) more prey than individual parasites. Great care must be taken in predator selection. Some predators are less host-specific than parasites and care must be taken to ensure that other beneficial species are not eaten when the pest species has been eliminated.

Insect predators occur in many orders including Coleoptera (beetles), Neuroptera (lacewings), Diptera (flies) and Hemiptera (bugs). Predaceous coleopteran families include the Coccinellidae (the well-known lady-beetles), the adults and larvae of which prey on homopterans, especially scales and aphids. Members of the Neuroptera, such as green and brown lacewings, are common aphid and whitefly predators, in both the adult and larval forms. Economically important members of the Diptera include the Calliphoridae, Sarcophagidae, Anthomyiidae and Bombyliidae, all of which contain some important grasshopper or locust egg parasites, and the Syrphidae whose grub-like larvae are voracious aphid predators. Although the Hemiptera contains many plant-feeders, a substantial number of species, in several families, have evolved to become predaceous. These include some anthocorids, predators of leafhoppers and lepidopteran larvae and eggs.

4.5.2 Techniques of control

The need for pest control commonly arises because a pest has been imported. Many biological control measures simply involve reuniting the natural enemies, from abroad, with the pest in its new environment. In the case of native pests the situation is more complex since the native enemies are clearly not performing satisfactorily. Cultural controls may help by improving conditions for existing enemies. For example, the aphid predators, green lacewings, *Chrysopa* sp., were attracted to Californian fields by spraying a cheap protein hydrolysate that mimicked aphid honeydew. Alternatively, predators and parasites of closely-related foreign pests may be imported. Inoculative or inundative releases of natural enemies may be made. Inoculative releases involve small numbers of predators/parasites where control is expected from subsequent generations, not from the release itself. Inundative releases are a liberation of large numbers of natural enemies, literally a biological insecticide, that are often cultured under greenhouse conditions. Mass releases of *Trichogramma* sp. have been tried many times, not always with success. However, the inundative release of two parasitoids to control black scale, *Saissetia oleae*, and citrus red scale, *Aonidiella aurantii*, in the Filmore Citrus Protective district of California, provided effective control for under $20/ha.

4.5.3 Desirable characteristics of biological control agents

Before the often expensive mass culture of natural enemies it is advisable to screen for beneficial attributes. These include:

(i) high searching capacity, even at low pest densities, to prevent emigration of enemies;

(ii) host-specificity; monospecific organisms 'track' host populations more effectively than polyphagous types, and do not 'switch' to other hosts;

(iii) high reproductive rate, influenced by development time, fecundity, number of generations per year and in some cases by parthenogenesis;

(iv) climatic adaptation to area of release;

(v) ease of laboratory rearing, especially on alternative food sources.

4.5.4 Famous successes of biological control

In 1972 it was estimated that 253 of the 356 biological control attempts worldwide had been completely, substantially or partially successful. Probably the 'grandaddy' of them all was the classic case of the vedalia beetle that was imported from Australia in the 1880s to control cottony cushion scale on Californian citrus (p. 13). Many of the greatest successes have been against sessile insects that cannot easily escape natural enemies. Some other examples are given below.

The coconut moth, Levuana iridescens, in Fiji

In the 1920s the most profitable Fijian industry, the copra industry, was threatened by coconut moths. These were endemic to the region and their larvae defoliated entire stands of coconut. A £5000 reward was quickly offered for remedies, but remained unclaimed. The solution was only forthcoming after hours of travel and search, throughout the Pacific, by several professional entomologists. Eventually, after an outbreak in Malaya of a related moth, *Artona catoxantha*, the parasite of this moth was laboriously shipped to Fiji. Within two years this parasitic tachinid, *Ptychomyia remota* had provided such dramatic control that scarcely a coconut moth could be found. Environmental conditions for this project, and for others on tropical islands, were favourable in two ways. Firstly, the operation involved an island with a limited pest population and, secondly, the mild climate permitted overlapping pest generations providing a continous, uninterrupted supply of hosts.

The greenhouse whitefly, Trialeurodes vaporarium, in Britain

In 1926 a gardener in Britain noticed that troublesome whitefly scales on cucumber plants had turned black. These were identified as parasitised by the small wasp, *Encarsia formosa* (Fig. 4.16). This wasp breeds rapidly, is parthenogenetic, produces several generations a year and can be used to achieve complete control at 24–27°C. In 1929, a special greenhouse was produced at Cheshunt Experimental Station to be used as an *Encarsia* 'factory'. In 1930, over 1 000 000 parasitised scales were developed on tomato leaves, and supplied for control to nearly 60 growers. This method proved effective, providing over 50% parasitisation was achieved. It was used until 1946 when DDT and smoke stopped biological control for nearly 20 years. The subsequent appearance of resistant strains has resulted in renewed interest in biological control.

The walnut aphid, Chromaphis juglandicola, in California

Walnut, *Juglands regia*, and the walnut aphid have become widely distributed, from their original habitat in south-eastern Europe and China. In California, the aphids became so abundant that honeydew,

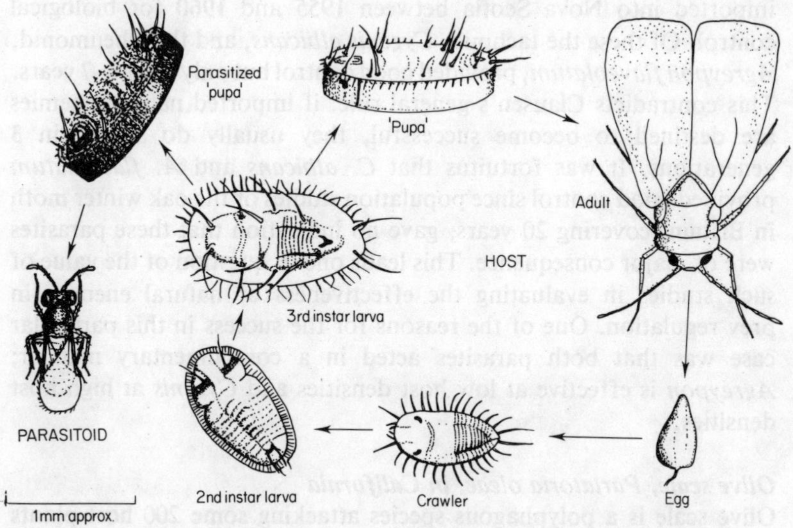

Fig 4.16 Life-cycle of the greenhouse whitefly, *Trialeurodes vaporariorum*, and its parasitoid, *Encarsia formosa*. The female *E. formosa* lays its eggs inside the third-instar larva. The whitefly continues to develop but turns black at the pupal stage instead of the usual greenish-white. An adult parasitoid emerges from the pupal case.

and the sooty mould that grows on it, literally covered the trees. It was a widely-held belief that aphids were poor targets for biological control since their high reproductive rate and ability to reproduce at low, spring temperatures gave them an insurmountable advantage over natural enemies. A wide variety of insecticides had been used over the years with all the attendant problems of resistance and environmental contamination. In 1959 the parasitic wasp, *Trioxys pallidus*, was imported from French walnut groves. In the mild coastal areas the parasites quickly became established but in the hotter, more arid conditions of Central Valley, in the interior, there was little success. Later, a new strain of *T. pallidus* was imported from the hot dry central plateau of Iran and this produced spectacular results. Within two years, biological control was completely successful.

The winter moth, Operophtera brumata, in Canada

This pest was accidentally introduced into Nova Scotia from Europe in the 1930s. By the 1950s it occupied a third of the province and in a space of 10 years the larval damage in just two counties amounted to 26 000 cords of wood, worth about $2 million.

Out of 63 species known to attack *O. brumata* in Europe, six were imported into Nova Scotia between 1955 and 1960 for biological control. Of these the tachinid, *Cyzenis albicans*, and the icheumonid, *Agreypon flaveolatum*, provided good control but only after 6–7 years. This contradicts Clausen's general rule: if imported natural enemies are destined to become successful, they usually do so within 3 generations. It was fortuitus that *C. albicans* and *A. flaveolatum* provided good control since population studies of the oak winter moth in Britain, covering 20 years, gave no indication that these parasites were of major consequence. This leads one to question of the value of such studies in evaluating the effectiveness of natural enemies in prey regulation. One of the reasons for the success in this particular case was that both parasites acted in a complimentary manner; *Agreypon* is effective at low host densities and *Cyzenis* at high host densities.

Olive scale, Parlatoria oleae, in California

Olive scale is a polyphagous species attacking some 200 host plants including olive, peach, apricot, plum and almond. Soon after its discovery in the US in 1934, the scale became a severe pest, and it remained so until a biological control programme was established in 1948. Olive scale, although known from the Mediterranean region for centuries, was believed to be native to India and Pakistan, so searches for natural enemies covered a large area. At least three different strains of *Aphytis maculicornis* were imported and released; from Egypt (a thelytokous strain), Spain (a deuterotokous strain) and Iran and India (an arrhenotokous strain). It was soon evident that only the 'Persian' variety would be effective but even populations of this strain were greatly reduced by the summer heat. Later, during a search for enemies of another pest, citrus red scale, the parasite *Coccophagoides utilis* was found in olive scales in west Pakistan. This parasite was shipped to California and provided sufficient extra control to render olive scales no longer a problem. Again, two parasite species are acting in a complimentary manner: *Aphytis* is most effective in spring and *Coccophagoides* in summer.

The question of whether to release a single species of parasite or many species is one of the most debated issues of biological control today. The argument against multiple species releases is that competition may occur between the natural enemies, reducing their effectiveness. However, in practice, despite competition, nearly all multiple releases provide greater overall control than single species releases. This is illustrated by the control of *Dacus dorsalis*, the oriental fruit fly, in Hawaii in 1949 (Fig 4.17). Taxonomic neglect

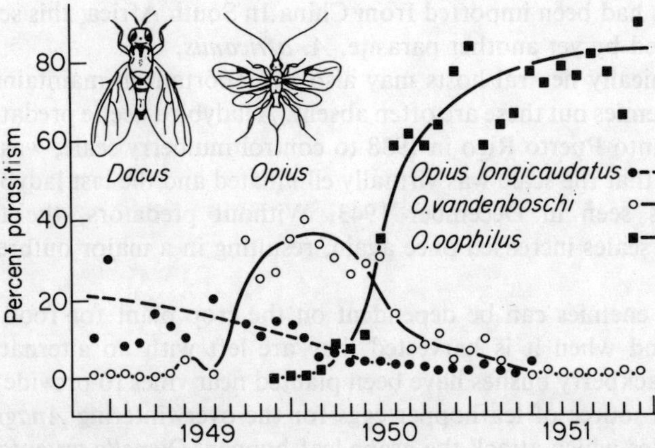

Fig 4.17 Introduction into Hawaii of three parasitoids of the fruit-fly *Dacus dorsalis* (Diptera: Tephritidae).

resulted in the simultaneous release of three species of parasite, *Opuis longicaudatus, O. vandenboschi* and *O. oophilus*, when the intention was to release only *O. longicaudatus*. First, *O. longicaudatus* became established, then *O. vandenboschi* and, finally, *O. oophilus* competitively displaced both species, maintaining a higher degree of control then either of its contemporaries.

4.5.5 Reasons for failure

Since the initial success against the cottony cushion scale in California over 150 species of insect pests have been completely or substantially controlled by introduced predators or parasites. Despite this gratifying record, many attempts in classical biological control meet with failure or only limited success. The reasons for failures or delays in success are many and varied. Probably the most important is poor adaptation to different climates. Small variations in temperature and humidity are critical to the success of most species, as illustrated by the walnut aphid and olive scale parasites (pp. 64 and 66).

Aphytis lingnanensis, imported from China in 1947 to control the California red scale, *Aonidiella auranti*, provides control in the mild coastal regions but not in the hot-summer and cold-winter region of the interior where only *Aphytis melinus* is effective. It has also now been established that *A. lingnanensis*, could have been collected, at much less expense, from Southern Texas, only 480 km away from the California citrus groves! In the past, the movement of Spanish galleons from the Orient to Mexico had facilitated the establishment of *A. lingnanensis* in the New World, but it was not detected until after the species had been imported from China. In South Africa, this scale is controlled by yet another parasite, *A. africanus*.

Economically neutral hosts may also be important in maintaining natural enemies but these are often absent. A ladybird beetle predator, imported into Puerto Rico in 1938 to control mulberry scale, was so successful that the scale was virtually eliminated and the last ladybird beetle was seen in December 1943. Without predators, the few remaining scales increased once again, resulting in a major outbreak in 1955.

Natural enemies can be dependent on the crop plant for food or habitat, and when it is harvested they are left with no alternative source. Blackberry bushes have been planted near vines to provide an alternative source of leafhopper eggs for the overwintering *Anagrus* egg parasites which attack the grape leaf-hopper, *Dicnella cruentata*, in summer (Section 4.1).

Another main reason for failure of biological control programmes is the 'shotgun' approach: parasites of closely related species from around the world are introduced in the hope of control. For example, by 1955, 38 species of natural enemy had been introduced into California against the black scale, *Saissetia oleae*. Parasites from such widely scattered places as Brazil and Tasmania were tested when, in fact, the genus *Saissetia* is native so South Africa. It is hardly surprising that the only one to have any important effect was a South African species. However, simple broad-spectrum introductions can be found in programmes involving such pests as the gypsy moth, *Lymantria dispar*, the brown-tail moth, *Nygmia phaeorrhoea*, the oriental fruit moth, *Cydia molesta*, and the European corn borer, *Ostrinia nubilalis*.

4.6 'Future' control techniques

There are many unusual techniques for controlling pests which have met with success but only on a few occasions. Current research is trying to thrust these control measures into the mainstream of pest management. However, considerable work still needs to be done to enable such methods to become widely-accepted techniques.

4.6.1 Microbial control

Over 1000 pathogenic bacteria, viruses, fungi, protozoa and nematodes have been described as specific pathogens of insects. Most pathogens need to be ingested but some act as contact agents, such as the fungi *Beauvaria* (a caterpillar pathogen) and *Entomophthora* (an aphid pathogen).

The most useful control agents so far have undoubtedly been the *Bacillus* bacteria and certain viruses. *B. popillae*, effective aganist Japanese beetle larvae, forms toxic protein crystals inside its host at the time of sporulation, turning the blood into a milky liquid. *B. thuringiensis* is the only microbial control agent registered in the USA. It is used against caterpillars on many crops, but, due to its short life-cycle, repeated applications are necessary.

Of the viruses, the nuclear polyhedrals attack a wide range of caterpillars, most of which then develop the habit of climbing to vegetation tips before they die – perfect for the release of virus spores. Granulosis viruses have also been used with success in Canada against the European sawfly, *Diprion hercyniae*. Microbial control has many advantages including:

(i) no toxic residue;
(ii) high specificity;
(iii) low dosage requirement (agent will multiply on its own);
(iv) insecticide-like application methods;
(v) resistance is unlikely.

However,a few disadvantages have restricted its general uses so far:

(i) careful timing is needed to apply at the host's infective stages;
(ii) pathogens do not disperse well on their own;
(iii) rotting cadavers are unacceptable in produce.

4.6.2 Insect pheromones (see also section 4.2.2)

Pheromones (from the Greek 'pherein', to carry, and 'horman', to excite) are chemicals that are released by insects and carried in the air, and which can illicit sexual behaviour, aggregation, dispersal, ovipostion or alarm in other individuals. Aphids rapidly disperse on detection of Trans-β-farnesene, the aphid alarm pheromone. The most useful pheromones today are those illiciting sexual behaviour and attracting individuals. Sex pheromones are secreted by a variety of abdominal or wing glands and detected by the antennae.

Synthetic analogs of many major pest pheromones are now available and are often used to bait traps. After 30 years, and one incorrect chemical determination, the sex pheromone of the gypsy moth was isolated and synthesised in the laboratory as 'Dispalure'. Despite intense trapping efforts the moth is still spreading; an indication that the technique is indeed one for the future. Mass release of pheromones in the field, for disruption of normal mating, has probably been the most successful development so far in this area.

4.6.3 Radiation sterilisation

The development of this method is largely due to the highly-successful campaign to control the screw worm, *Cochliomyia hominovorax*, a pest of livestock. On the small Caribbean island of Curaçao, and in Florida, sterile males were released to compete with wild males. Eventually, the huge releases of sterile males reduced the chances of successful mating and viable progeny to such a degree that the screw worm was eradicated. Attention then shifted to the south and south-western USA, an overwintering stronghold for flies. The sterile release plan began again. Male flies could be sterilised by as little as 25 Gy (compared to the 600 Gy needed for some insects, such as pink bollworms). A 5000 roentgen cobalt bomb was exploded, in the laboratory, to sterilise the flies (1 roentgen is roughly equivalent to 0.01 Gy). The gamma radiation produced proved very effective: it has more penetrating power than alpha or beta waves but it much cheaper to produce than X-rays. Following the releases, eradication was complete in eight weeks. Other success stories using this method include control of the oriental fruit fly, *Dacus dorsalis*, in Guam, and of the mediterranean fruit fly, *Ceratitis capitata*, in California. However there are many prerequisites for success:

(i) females must mate only once;
(ii) sterilised males must compete favourably with normal males;
(iii) sterilised males must disperse radially;
(iv) for total eradication, release must be made when population density is low (Table 4.6).

Even under these conditions, success can not be guaranteed. In the screw worm campaign of the south-western USA, 50 tonnes of blood and meat and 150 million sterilised pupae had to be released along the Mexican border to prevent the spread of screw worm from Mexico, after it had been eradicated in the USA. Seven years after eradication in the USA,90 000 new cases of screw worm were reported indicating that the technique had yet to be perfected.

4.6.4 Chemosterilants

Chemosterilisation works in a similar manner to radiation sterilisation in that it reduces the numbers in the next generation. It differs is that it does not entail the laborious processes of rearing, irradiating and then releasing insects. Despite these advantages, examples of control using chemosterilants are scarce.

Four classes of chemosterilants are recognised: phosphorus amides; triazines; antimetabolites and alkylating agents. Antimetabolites are structurally similar to biologically active metabolites (such as purines and pyrimidines) and are incorporated into nucleic acid. Alkylating

Table 4.6 Theoretical population decline in each subsequent generation when a constant number of sterile males is released among a natural population of 1 million females and 1 million males.

Generation	Number of virgin females in the area	Number of sterile males released each generation	Ratio of sterile to fertile males competing for virgin female	Percentage of females mated to sterile males	Theoretical population of fertile females each subsequent generation
F_1	1 000 000	2 000 000	2:1	66.7	333 333
F_2	333 333	2 000 000	6:1	85.7	47 619
F_3	47 619	2 000 000	42:1	97.7	1 107
F_4	1 107	2 000 000	1 807:1	99.95	Less than 1

agents comprise the biggest group of sterilants. Many contain aziridinyl groups,

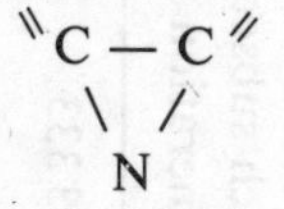

the action of which may prevent DNA synthesis. These agents are highly reactive with proteins and nucleic acids, replacing hydrogen with an alkyl group (e.g., CH_3 or C_2H_5). Their effects on insects are similar to those of irradiation, hence their common name, 'radiomimetics'. They cause meiotic disturbances that are most intense where cell division is taking place resulting in lethal mutations and zygotes (fertilized eggs) which cannot survive to maturity. The potential danger to humans has restricted the use of these compounds.

4.6.5 Insect growth regulators

Insect growth regulators (IGRs) are often called 'third generation pesticides', stomach and contact poisons being the first and the second generations respectively. The discovery of IGRs was possibly one of the most serendipitous finds in Science. Cultures of a European bug, *Pyrrhocoris apterus*, that were shipped to the USA, always developed an extra instar instead of moulting to adulthood. Similar European cultures always emerged normally. Eventually, the difference was traced to the paper towels lining the bugs cages. In the USA they were made from balsam fir, *Abies balsamea*, a species not used in Europe. The paper was found to contain a close analog of juvenile hormone which, when ingested, increased the level of juvenile hormone in the insects' blood preventing the final moult to adulthood. Immediately, juvenile hormone (JH) which prevents metamorphosis, and ecdysone, the insect moulting hormone, became the subjects of intense speculation regarding their potential in pest control. Many plants are known to contain ecdysone derivatives and these often play a key role in the plant-insect relationship. Ecdysone derivatives have been extracted from plants and applied to insects at the last nymphal or larval stage. This produces giant nymphs or larvae, incapable of reproduction.

The most promising IGRs now on the market are Altosid ® and Dimilin ® both marketed by the Zoecon corporation. They are used against 4th instar mosquito larvae at 0.05 kg/ha, and they prevent adult emergence by inhibiting chitin synthesis. For a long time it was thought that no species could develop immunity because JH is such an integral part of insect physiology. However, this bubble has just been burst by the development of resistance in houseflies and mosquitoes in the laboratory. Resistance develops in much the same way as pesticide resistance.

4.6.6 Competitive displacement

This involves introducing an innocuous, highly-successful, competitive species, to displace an indigenous species. A good example is provided by the use of African dung beetles in Australia. Cattle are not native to Australia but were brought in by Europeans in the 18th century. More than 300 million dung pats are deposited every day and these need to be disposed of. Unburied dung is a problem because two major pests, the blood-sucking buffalo fly, *Haematobia irritans exigua*, and the bushfly, *Musca vetustissima*, breed in it. Pats also stimulate the growth of rank grass on which cattle will not feed. Although there are 250 native species of dung beetle, they are adapted to deal with the small, coarse pellets of indigenous marsupials. African dung beetles were imported to cope with the cattle dung.

Of the 1 800 species of dung beetle in Africa, so far 44 have been sent to Australia for control purposes. Most of these have become established and have spread. The beetles either bury the dung in tunnels or roll it away in balls, and then lay their own eggs in it. Research has shown that beetles which bury cow pats within 48 hours can prevent an average of 96% of the flies from maturing and 85% of parasitic cattle-infesting worms from developing. In addition, some beetles carry mites of the family Macrochelidae from one pat to another (phoresy) and many of these are effective predators of fly eggs and larvae.

5 Integrated pest management

The survey of pest management techniques, presented in the last chapter, dealt largely with individual methods of control. This is a convenient method of documentation but also reflects the discrete way in which these methods have been employed. The modern concept in pest management is that of integrated control in which many control techniques are used, simultaneously, to provide the best possible control. This is a noble principle but one that is difficult to achieve in practice owing to the many variables involved. Growers generate heavy pressure for quick answers to pest problems, so the more long-term integrated pest management strategies are passed-over in favour of short-term methods. This is a great pity since there are well-documented examples that show how a combination of control techniques can have multiplicative, not just additive, effects.

The LC_{50} for *Myzus persicae* to parathion, on partially resistant chrysanthemums, is less than half that on susceptible varieties. Similarly, the survival of leafhoppers exposed to dimethoate was reduced by a third when they were on resistant, as opposed to susceptible, cowpea varieties. Incorporating an attractant with an insecticide in citrus sprays attracts fruit flies toward the insecticide. If only the lower half of the trees are sprayed any pests remaining in the upper half are reduced by the now unnaturally high proportion of natural enemies.

A DDT band can be painted round coffee tree bases and whenever levels of pests (particularly caterpillars) increase, pyrethrum can be applied at minimal knock-down levels. As the affected insects recover, most pest parasities and predators can *fly* back into the tree, whereas many pests, particularly the caterpillars, have to *walk* across the DDT band which eventually kills them.

Despite the difficulties of implementing integrated pest management schemes, a more holistic approach to the problems of insect pest control is now being adopted for some control programmes. Some of the best examples are given below.

5.1 Alfalfa in California

It was in California that the theories of integrated pest managment were initiated, and it is therefore appropriate to consider the case of alfalfa in California, as one of the first crops where a holistic approach was used .

Alfalfa is a major foodsource for livestock and poultry and an important rotation crop, enhancing nitrogen and organic matter. In California alone it covers 600 000 ha. One of the first steps was to establish economic injury thresholds for its major pest, *Colias eurytheme*, the alfalfa caterpillar. Thresholds were set at 10 *healthy* larvae per sweep. Only healthy larvae were counted, since it was recognised that the parasitic wasp, *Apanteles medicaginis*, and a nuclear polyhydrosis virus were often able to exert considerable natural control. Having established economic thresholds, pesticide application could be kept to a minimum. In 1954, the devastating spotted alfalfa aphid, *Therioaphis trifolii*, arrived and soon threatened to wipe-out the alfalfa industry, with $10.6 million losses in 1957 alone. However, successful modifications to the existing IPM programme reduced these losses to $1.7 by 1958 – only one year later. Elements of the successful programme included:

(i) reduced spraying to enhance the activity of ladybird beetle predators (*Hippodamia* sp.);
(ii) introduction of three species of hymenopterous parasite from Europe and the Near East;
(iii) strip havesting to provide refuges for these natural enemies;
(iv) timely irrigation procedures to enhance virus action;
(v) introduction of a more selective insecticide, demeton (Systox ®), which killed more aphids but fewer natural enemies;
(vi) introduction of new resistant varieties of alfalfa which finally removed the need for insecticides altogether.

5.2 Apples in Nova Scotia

This is often regarded as the 'grandaddy' of all deciduous fruit-tree crop control programmes. Pests that ruin the appearance of a crop have economic injury levels of zero, because a product on the market must be attractive. This necessitates heavy insecticide dosages very early on, and the establishment of a 'pesticide treadmill' which is easy to start but difficult to stop. If natural enemies fall victim to the pesticides more species can reach pest status. The codling moth, *Laspeyresia pomonella*, (the common worm in the apple) and the eye-spotted bud moth, *Spilonata ocellana*, were quickly joined in pest status, early in the 1940s, by previous non-pests such as oystershell scale, *Lepidosaphes ulmi*, European red mite, *Panonychus ulmi*, and the grey-banded leaf-roller, *Argyrotaenia mariana*. The situation was eventually remedied by:

(i) the use of copper-based fungicides instead of sulphur-based ones which had killed off the oystershell scale enemies;

(ii) the introduction of the selective stomach insecticide, ryania (a 'natural' insecticide from stems and roots of the S. American shrub, *Ryania speciosa*) aganist codling moths and other pests applied at very low densities to allow European red mite natural-enemy complexes to increase;

(iii) the introduction of two parasites from Europe to control the winter moth, *Operophtera brumata*, which appeared in the 1930s;

(iv) cultural controls, especially sanitation (the removal of wild hosts and fallen fruit) to control apple maggot, *Rhagoletis pomonella*.

5.3 Tobacco in North Carolina

Here the major pests are dealt with sequentially. Tobacco budworm, *Heliothis virescens*, is an early season pest inflicting damage on developing leaf buds. Reduced spraying allowed natural enemies – the stilt bug, *Jalysus spinosus*, and the parasitic wasp, *Campoletis peridstinctus* – to exercise control. The tobacco hornworm, *Manduca sexta*, feeds on larger leaves and therefore may be a pest throughout the season. Reduced sprayings allowed its natural enemies – the stilt bug wasp, *Apanteles congregatus*, and the tachinid fly, *Winthemia manduca* – to exert control, so that often only one spray per season, or less, was needed. Flea beetles, *Epitrix hirtipennis*, cause only cosmetic damage to the leaves. However, growers still tend to pride themselves on unblemished leaves and to convince them not to spray is not easy. Spraying can kill aphid natural enemies, causing pest outbreaks which normally do not occur. Nematodes, primarily root-knot nematodes, are reduced by resistant varieties and by rotation. Resistant varieties are also available for the important microbial pests: tobacco mosaic virus, bacterial wilt and 'black shank' fungus. Dipping hands in milk solution during planting often reduces mosaic virus.

Since tobacco is grown for its leaves, the tops, seeds, fruit and flowers are superfluous and are usually removed. This stimulates increased vegetative growth and has the advantage of removing suitable oviposition sites for hornworm populations. Post-harvest stalk destruction is also important because stalks are a food source for developing pest larvae which will then not be able to survive the winter.

5.4 Cotton in California

Production of cotton is ancient, dating back at least 5000 years. Over 1300 insect species throughout the world have evolved to feed on this crop. This is the king of the 'pesticide treadmill' crops, with the cotton boll weevil, *Anthonomus grandis*, (Fig 5.1) being the world's most serious pest, damaging about 11% of the crop in the USA alone.

In California, the key pests are the lygus bug, *Lygus hesperus*, which feeds on the fruiting bolls causing them to shrivel and drop off, and delaying fruit (boll) formation, and the cotton bollworm, *Heliothis zea*, which attacks the bolls. The bollworm is most feared because feeding scars allow the penetration of rot-producing microorganisms.

Cotton is perennial, but most crops are replaced each year. In some regions, old plants are cut off ('ratooned') several centimetres above the ground and allowed to regrow. This is favoured by growers where water for irrigation, and labour for planting are scarce. It enables bollworms to survive to the next year and so is illegal in many areas of the USA.

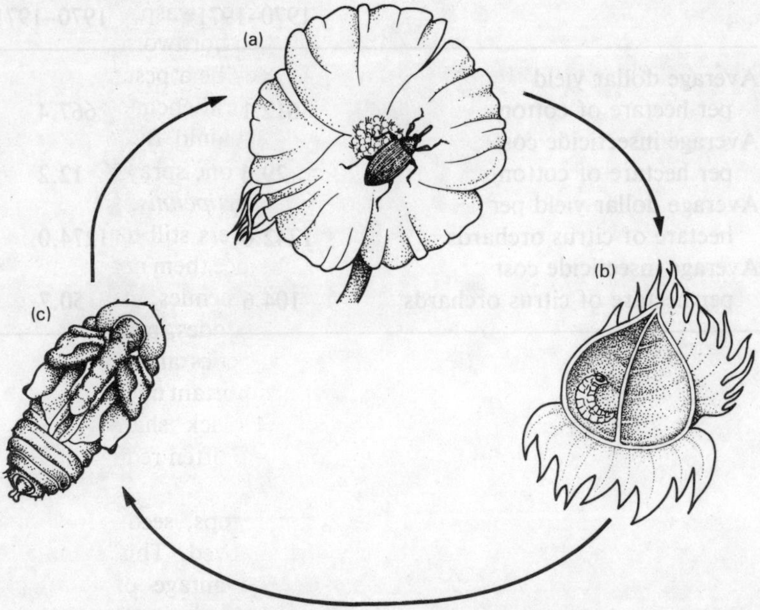

Fig 5.1 Life-cycle of the boll weevil, *Anthonomus grandis*: (a) adult on flower; (b) larva in boll, and (c) pupa.

Two other minor pests are the beet armyworm, *Spodoptera exigua*, and the cabbage looper, *Trichoplusia ni*.

In California, pest 'status' was removed from these insects by two major breakthroughs.

(i) Use of chemical sprays was reduced, since the bollworm had attained pest status as a result of extinction of natural enemies by chemicals. 41% of bollworm eggs are eaten by predators in 24 hours in unsprayed areas. Certain insecticides (methyl parathion and carbaryl) were found to have a direct detrimental effect on the plant and application of these was also stopped.

(ii) More accurate field scouting showed that sprays against the lygus bug need only be applied during the budding season. This, in turn, allowed the natural enemies of the two minor pest species to proliferate and exert control. *Trichogramma* sp. destroyed up to 50% of cabbage looper eggs.

Table 5.1 Benefits to cotton and citrus groves in central California (San Joaquin Valley) of employing IPM strategies.

	Non-users 1970–1971	Users 1970–1971
Average dollar yield per hectare of cotton	612.1	667.4
Average insecticide cost per hectare of cotton	29.6	12.2
Average dollar yield per hectare of citrus orchards	1242.0	1274.0
Average insecticide cost per hectare of citrus orchards	104.6	50.7

6 Insect herbivores as control agents for weeds

There are many similarities between the control of weeds and the control of other pests, and insects themselves feature prominently in weed control. Losses to weeds are immense: in 1975 in the USA alone, $5 billion was lost in crop production because of competing weeds, and a further $5 billion was spent trying to prevent that very loss. Again, most problems arise from the importation of weeds without their full compliment of natural enemies. Of 693 species classified as weeds, in California, 63% are aliens and, in general, these are the most serious pests. The techniques of weed control are many and varied. They include many physical, mechanical and cultural technqiues such as pulling, hoeing, mowing, burning, crop rotation and flooding. However, the two main techniques involve the use of herbicides and of biological control.

6.1 Herbicides

The application of herbicides is often fraught with difficulties, especially since nearby crops are often also susceptible to broad spectrum herbicides. The cost of developing a herbicide for one particular species of weed would be prohibitive; the returns would not be sufficient to cover the costs (see section 4.4.4).

However, the herbicide industry really began to accelerate in 1945 with the discovery of at least some selective phytotoxicity in 2,4–D (2,4–dichlorophenoxy acetic acid) which was toxic to many broad-leaved dicotyledonous weeds, but not to many monocotyledonous grass crops. In 1950 there were only 15 primary herbicides available; in 1970 there were 180. More than 60 million hectares, or 50% of the total harvested crop area, was sprayed with herbicide in 1973.

Herbicides may be selective, such as 2,4–D, or non-selective, such as paraquat or amitrole, which kill all vegetation. They may act directly as contact poisons, killing the foliage on which they are sprayed (for example, paraquat, diquat, dinoseb, propanil and bromoxynil) or they may be translocated through the plant and accumulate in the actively growing areas, such as the meristems. Dalapon, dicamba and picloram are common translocated herbicides, as are the phenoxy or 'hormone' herbicides, 2,4–D, 2,4,5–T, silvex and MCPA which act as growth stimulants, causing the plant to literally grow itself to death. Herbicides that are applied to the soil before or soon after planting, may be translocated or act directly on the root. These include atrazine, benefin, bromacil, chloramben, molinate, monwon, nitralin, prometryne, propham, cycloate, diwon, simazine, endothall, EPTC and trifluralin. A very few herbicides such as dinoseb, dicamba, and picloram may be applied to soil or foliage.

6.2 Biological control

In biological control of insect pests the agents usually kill the prey directly. In biological control of weeds the agent may:
(i) directly eat and kill the host;
(ii) weaken or stress it, lowering resistance to disease, other herbivores or competitors;
(iii) impair the reproductive potential by eating the flowers, fruits or seeds.

The most crucial aspect of weed control by insect herbivores involves rigid and extensive host-plant choice experiments to ensure that agents do not switch to alternative hosts, such as crops, once the weed is removed and starvation sets in. Important weeds such as prickly pears and klamath weed are so taxonomically isolated from any economically important plant that they are good candidates for control (p. 81). The blackberry, *Rubus fruticosus*, an important weed in New Zealand, is a member of the Rosaceae, which also contains apples, pears, raspberries and a host of ornamentals. In this situation there seems to be little prospect of blackberry control by introduced biocontrol agents.

Regarding introduced agents becoming pests, there appear to be only a few isolated cases; examples are *Thecla echion* attacking eggplant (aubergine) in Hawaii and certain insects attacking spineless cacti in South Africa.

Of the 600 attempts at biological control of approximately 84 weeds, insects were involved in 174. Amongst the notable successes is control of the lantana weed, *Lantana camara*, in Hawaii in 1900 by tortricid moth caterpillars, lycaenid butterfly larvae and *Agromyza* fly larvae. Another success is the elimination of alligator weed in Florida inland water-ways by the leafbeetle, *Agasicles hygrophila*. Several other successes are described below.

Tansy ragwort, *Senecio jacobea*, is a pest species native to Europe but is now established in many areas of the world. It is poisonous to livestock because it causes toxic alkaloids to accumulate in the liver. Since 1927 the cinnabar moth, *Tyria jacobaeae*, and the fly, *Hylemya seneciella*, have been regarded as possible biocontrol agents; the moth defoliates the plant and the fly larva attacks the flower-heads. For control in New Zealand, over ¼ million moth pupae were imported from England in 1927–1930, followed by ½ million fly larvae in 1937. The fly became established but the moth did not. This was possibly because of the attentions of native parastoids of the related magpie moth. In Australia, predation by the scorpion fly, *Harpobittacus* sp., also killed the moth and the fly did not become established. However, in Canada and the USA, shipments of moths soon began to effect control, defoiliating the plants and allowing the cold winters to finish the job. This project is typical of many being carried out against weeds. Success has not been spectacular: it has been achieved in some areas by careful research; in others it has not been achieved at all. There are, however, one or two cases of spectacular control that are the pride and joy of the biocontrol specialists.

In 1788, when the First Fleet arrived at the site of what is now Sydney, the ships carried a mixed cargo of convicts, soldiers, officials and a cactus, the smooth pear tree, *Opuntia monacantha*, which Governor Phillip had collected from Rio de Janeiro with the intention of starting a cochineal industry. The cochineal industry never became established but the cactus did, along with several other *Opuntia* species (possibly escaped ornamentals). By 1900 over 40 000 km^2 were infested and further land was lost at the rate of almost 10 000 km^2 each year. At the height of the infestation, in 1925, 243 000 km^2 in Queensland and New South Wales was lost. The introduction in 1913 of *Dactylopius indicus*, a mealybug, gave satisfactory control of *O. monacantha*, but there was no respite from the other species, particularly *O. inermis* and *O. stricta*. In 1920, the Commonwealth Prickly Pear Board was set up and soon dispatched agents to the Americas to search out arthropod enemies of cacti. In 1925, the search ended with the shipment of the South American moth, *Cactoblastis cactorum*, from Argentina. By 1930, the voracious larvae had destroyed whole stands of prickly pear which eventually precipitated a moth food shortage, a decrease in the population of moths and increase in the numbers of cacti. These oscillations were short-lived, however, and by 1935 only isolated patches of cacti remained.

Klamath weed, *Hypericum perforatum*, also known as St. Johnswort was originally a perennial of Europe and Asia but has now invaded many semi-arid to sub-humid areas of the world, including Australia, New Zealand, Chile, Argentina, South Africa and North

America. It was first found in the 1900s in the USA, in Northern California near the Klamath river, hence the common name. By 1940 it covered more than 800 000 ha in the USA and had displaced valuable forage crops. Moreover, klamath weed contains hypericin, a compound which causes blindness and skin irritation in cattle. From the South of France, the beetles *Chrysolina hyperici, C. quadrigemina* and *Agrilus hyperici* were exported to Australia and provided encouraging control. After watching this project with extreme interest, growers in California imported these species in 1944. *A. hyperici* was lost but, within three years, the two *Chrysolina* bettles, in particular *C. quadrigemina*, had begun to exert their influence, despite producing only one generation a year. By 1956, most infected areas had been controlled and enthusiastic growers erected a monument to the successful beetles. Control in California had been more successful thăn Australia, and beetles were sent to Washington, Oregon, Idaho and Montana, where they also became established. In Canada, however, results were disappointing, perhaps owing to the vagories of the weather. Control that was achieved was provided largely by *C. hyperici*.

7 Insects as disease vectors

Insect pests of man and higher animals fall into two categories. The first consists of those that are a nuisance and includes cockroaches, some ants and bees, horseflies and deerflies (Tabanidae), the bedbug, *Cimex lechularius*, and the pubic louse, *Phthirus pubis*. The second category consists of more serious pests: those which serve as vectors of diseases, causing illness and death (Table 7.1). Of course a few insects may fit into both categories (e.g. *Musca domestica*, a vector of trachoma virus and bacterial dysentries in many parts of the world) but the distinction is still useful. Economic injury levels and economic thresholds are very much closer to the general equilibrium position for insect vectors of disease than for those pests which are merely a nuisance.

For a long time after the development of DDT, lindane and other modern insecticides, it appeared as though vector-borne diseases would soon be a thing of the past. Insecticides still remain as the single most important control measure, but the inevitable development of resistance has slowly (but severely) begun to render control programmes ineffective. The basis of disease prevention is to apply insecticide precisely at the site of host-vector contact, and to employ other control techniques to reduce pest densities. Several examples of control attempts against contemporary vectors are given below.

The mass delousing of the inhabitants of Naples in 1943, by application of DDT powder to clothing, virtually eliminated the human louse, *Pediculus humanus humanus*, and the concomitant threat of epidemic typhus, *Rickettsia prowazekii*. Dusting is still carried out all over the world but the dusts have changed from DDT to lindane, pyrethrins and carbaryl, as resistance has repeatedly reared its head.

Disease transmission may take several forms. Simple mechanical transmission involves the direct transference of disease from the source, e.g. rotting meat, to the host, on or in a vector's body. In this way, a fly may crawl over a dead animal and then spread bacteria over human food. No multiplication or development of pathogens occurs within the vector. In contrast to this rather casual manner of transference, in cyclical transmission, multiplication and/or development of the pathogen occurs within the vector. This is well developed and highly efficient in malaria. The parasites multiply within

Table 7.1 Some major vector-borne diseases

Disease	Vectors
Malaria (*Plasmodium* sp.)	*Anopheles* mosquitoes
Filariasis (*Wuchereria bancrofti*)	Anopheline and culicine mosquitoes
Yellow fever	Mostly *Aedes* mosquitoes
Dengue fever	*Ae. aegypti and Ae. albopictus*
Encephalitis viruses	Culicine mosquitoes and ixodid (hard) ticks
River blindness (*Onchocerca volvulus*)	Simuliid blackflies
Sleeping sickness (*Trypanosoma* sp.)	Tsetse flies
Chagas' disease (*Trypanosoma cruzi*)	Triatomine bugs
Leishmaniasis (*Leishmania* sp.)	Phlebotomid sandflies
Endemic typhus (*Rickettsia prowazeki*)	Body lice
Endemic typhus (*Rickettsia mooseri*)	Rodent fleas
Plague (*Yersinia pestis*)	Rodent fleas

Anopheles mosquitoes and undergo sexual reproduction, producing thousands of infective sporozoites which accumulate in the salivary glands, ready to be released during the next feed.

Sometimes other animals harbour human diseases, acting as pools of infection. Zoonosis is the term applied to a disease that can be transmitted in this way. Examples include yellow fever from monkeys, Chagas' disease from rodents and sleeping sickness from game and cattle.

7.1 Mosquitoes

At least 15 species of malaria vector are resistant to DDT and 36 to dieldrin and lindane – the three insecticides on which the first eradication programmes were based. In India, prior to 1947 when residual house spraying started, there were about 75 million malaria cases annually. By 1965 there were only 100 185 cases and malaria had almost been eradicated. Thereafter deterioration set in and by 1977 there were 7–10 million people with malaria. The exact causes of this alarming reversal include, of course, the development of resistance, but also socio-economic changes resulting in decreased control budgets and poor entomological surveillance. Other mosquito-borne diseases that are increasing include filariasis, which can lead to such gross deformities as elephantiasis, yellow fever and dengue.

Despite the shortcomings of pesticides, simple drainage, diking, ditching and filling, have always severely reduced larval numbers providing good control in areas as diverse as the Panama Canal (1904–7), Pontine Marsh, Italy (1930s) and the Nile flood plain. Larviciding is used in conjuction with these methods but often pollutes the aquatic environment. Biological control by natural enemies seems to have taken a back seat for the moment. Many potential fish predators are highly susceptible to organochlorine larvicides. The most successful predator is the mosquito fish, *Gambusia affinis*, which has been distributed throughout the warmer parts of the world and is a voracious predator. It is said that six *Gambusia* can completely control mosquitoes in a 5–10 m^2 pool. Amongst the invertebrates, the predatory mosquito, *Toxorhynchites*, is the most promising candidate, especially as it can reach otherwise inaccessible breeding places.

7.2 Blackflies

There are over 1 000 species of simulid blackfly and most of the females suck blood. Heavy swarms are often a great nuisance in North America and Europe. In Africa, several species are vectors of the filarial parasite, *Onchocerca volvulus*, the organism that causes human onchocerciasis, or river blindness. In the Volta river basin alone, 1 million out of a population of 10 million are infected, with at least 70 000 totally or partially blind. Often when blindness affects more than 5% of a community, the community migrates, leaving hectares of fertile land behind. Blackflies are characterised by breeding in fast-flowing rivers or streams, and by the ability to disperse over long distances – commonly more than 20 km and sometimes up to 200 km. However, adults are rarely encountered away from water, except when biting man or domestic animals. Most control programmes are directed against the larval stages and sprays are rarely used against the adults. Temephos or methoxychlor are fortunately partially effective in low doses against the larvae and, at such low application rates, the environment is not harmed.

7.3 Tsetse flies

Tsetse flies, *Glossina* spp., have attained notoriety because they transmit pathogenic trypanosomes which cause human trypanosomiasis (sleeping sickness) and animal trypanosomiasis (nagana) in livestock. An estimated 35 million people risk getting sleeping sickness and about 10 000 new cases are reported annually. In animals, nagana can reduce growth rates by 40–50% and reduce milk yields to less than 1 litre per day. In Africa, there are 21 species of *Glossina*, distributed over 10 million km^2 between 15°N and 22°S.

Vector control is difficult and trypanocidal drugs still constitute the backbone of the trypanosomiasis control effort. Selective spraying (of adult resting places) by persistent insecticides provides some control of adults (see section 4.4.3.), whilst the maintenance of 5 km barrier zones, between game and domestic stock, in areas such as Zambia, greatly cuts down vector movements between source pools and domestic stock.

7.4 Triatomine bugs

Chagas' disease, *Trypanosoma cruzi*, or American trypanosomiasis is found in the Americas between 42°N and 45°S, where it infects 30 million people, of whom 5 to 8 million suffer permanent heart damage. It is transmitted by about 95 species of triatomine or reduvid bugs, the most important of which are *Triatoma* sp., *Rhodnius* sp. and *Panstrongylus* sp. It can be contracted by ingesting contaminated food or milk or by contact with the insects' faeces as well as from the insects when they feed on their hosts' blood. The best control measure would be to rehouse people in brick buildings instead of dilapidated mud and thatch dwellings, since Chagas' disease is primarily one of poverty: the bugs are secreted in the cracks and crevices of ramshackle buildings. Of course, rehousing is not economically expedient, and in practice spot treatments with insecticides, such as dieldrin, are commonly used. Immunologically there is no cure and no protection and humans remain infected throughout their lifetime.

8 Epilogue

Despite the impressive array of control techniques available to man, the rate of increase in food production in tropical developing countries is less than 1% annually, whilst the rate of population growth continues at an annual rate of 2.0–2.5%. Clearly we have far from solved all the problems. Half the clinical cases of disease in the world may be transmitted by insects. The losses of the world's major crops to insects continues to be high (Table 1.1). Even in the more developed areas, such as Europe and North America, crop losses to insects are 5% and 9% respectively, whilst in South America, Africa and Asia losses amount to 10%, 13% and 21% respectively. It is clear there are still many problems to solve and much research to be done.

It is often difficult, still, to distinguish certain insect damage from that caused by nematodes. It is, therefore, difficult to know whether to apply insecticides or nematicides, consequently no control measure may be taken. There are still few effective means of control for molecrickets in rangeland pasture grasses. Chemically treating huge areas of pasture is not economically viable; again no action is taken. Similarly, but in a lighter vein, the lovebug, *Plecia nearctica*, causes many complaints from tourists in Florida whose cars become smothered in copulating couples! It is possible that automobile exhaust fumes serve as lovebug attractants, but to identify the 'attractive' components of gasoline fuel has so far been impractical. A front ventilation screen is the most viable alternative – at the moment.

It is now obvious that pest management cannot stand alone, but must work with crop production to achieve satisfactory control of pests and good yields. Even then the demands of society may still be too great for the limited resources of the environment. Biologists are not magicians who can always conjure up an adequate food supply for the burgeoning human population.

Pest control receives great moral and financial support from the peoples of the world; it is a tragedy that its important counterpart, population control, is so hindered.

Glossary

Acetylcholine (ACh) Chemical transmitter of nerve and nerve-muscle impulses between nerves and across nerve-muscle juctions

Agroecosystem An agricultural area sufficiently large to permit long-term interactions of all the living organisms and their non-living environment.

Alkylating agent Highly active compounds that replace hydrogen atoms with alkyl groups, usually in cells undergoing division.

Ametabola Insects which do not undergo a metamorphosis.

Antennae (sing. antenna). Paired, sensory, segmented appendages found one on each side of the heads of insects and of some related forms.

Antibiosis The tendancy of a plant to resist insect injury often by injuring or destroying the insect.

Antimetabolite A chemical that is structurally similar to biologically active metabolites, and which may take their place in a biological reaction, to the detriment of the organism.

Appendage Any structure attached to a part of the body of an organism.

Attractant, insect A substance that lures insects. Usually classed as food, oviposition or sex attractants.

Autocide The destruction of a pest by itself or its own species.

Balance The state of an insect population in which large deviations from population oscillations do not occur.

Beneficial insect An insect that serves the best interests of man, e.g, insect pest predators and parasites, and pollinating insects.

Biological control Control of pests using natural enemies (usually predators and parasites).

Biomagnification The increase in concentration of a persistent pollutant along a food chain. Such pollutants include persistent organochlorine insecticides and their metabolites.

Biotic insecticide An insect pathogen that is applied in the same manner as a conventional insecticide to control pest species. Usually a microorganism is used (*see* microbial insecticide)

Biotype A population or group of individuals composed of a single genotype.

Broad-spectrum insecticide A non-selective insecticide that has about the same toxicity to most insects.

Carbamate insecticide One of a class of insecticides derived from carbamic acid.

Carrier (of a pesticide) An inert material that serves as a dilutent or vehicle for the active ingredient or toxicant.

Chemosterilant Chemical compounds that cause sterility or prevent effective reproduction.

Cholinesterase (ChE) An enzyme that is necessary for proper nerve function. It is inhibited or damaged by organophosphate or carbamate insecticides.

Cosmtic quality Aesthetic characteristics of a fruit or vegetable which have no relation to nutritional quality.

Coxa (pl. coxae) The basal segment of an insect's leg.

Cultural control A pest control method in which normal agronomic practices (tilling, planting, crop spacing, irrigating, harvesting, waste disposal and crop rotation) are altered so that the environment is less favourable for pests.

Dermal toxicity Toxicity of a material as a result of contact with the skin.

Diapause A physiological state of arrested development, generally resulting from physical stimuli, such as temperature and light. The ability to undergo diapause provides the insect a means of surviving unfavourable conditions.

Dilutent Component of a dust or spray that dilutes the active ingredient.

Drift Movement of airborne pesticide from the intended area of application.

Dynamic economic level A term used to denote a seasonally changing relationship, between a pest insect and its host plant, such that that the level of the pest population that is sufficient to cause damage varies according to the stage of growth and the season.

Ecdysone Hormone secreted by insects that is essential to the process of moulting.

Economic injury level The lowest insect pest density at which economic injury occurs.

Economic level The density of a pest population below which it fails to cause enough injury to the crops to justify the cost of the control effort.

Ecosystem The interacting system of all the living organisms of an area and their non-living environment.

Ectoparasite An organism living parasitically on the outside of another organism.

Elytra (sing. elytron) The hard, upper wings of beetles.

Emulsifier A surface-active substance used to stabilise a suspension of one liquid in another.

Emulsion Suspension of miniscule droplets of one liquid in another (e.g. oil in water).

Endoparasite A parasite feeding internally on host tissue.

Entomophagous Feeding on insects.

Fecundity Reproductive capacity.

Femur (pl. femora) The third (counting from the body) and usually heaviest segment of an insect's leg.

Formamidine insecticide A new insecticide with a new mode of action that is highly effective against insect eggs and mites.

Formulation The form in which a pesticide is sold for use (e.g. granules, dust, emulsion).

Fumigant A volatile material that forms vapours which destroy insects, pathogens and other pests.

Genetic control A pest control method which makes use of selected strains of the target species that possess genetic abnormalities. When released into the target population they mate with wild individuals and produce sterile offspring.

Grub The larva of a beetle.

Halteres Short-knobbed appendages present in place of the hind wings of true flies.

Holometabola Insects which undergo a complete metamorphosis.

Hormone A product of living cells that circulates in the animal or plant and produces a specific effect on cell activity away from its point of origin.

Host-plant resistance Inherited qualities of plants that influence the extent of insect damage.

Hydrolysis Chemical process of (in this case) pesticide breakdown or decomposition involving a splitting of the molecule and addition of a water molecule.

Inoculative releases The repeated release of relatively small numbers of a natural enemy for purposes of building-up a population over several generations.

Inundative release Release of large numbers of natural enemies to effect immediate high mortality in the pest population.

Insect growth regulator (IGR) Chemical substance which disrupts the action of insect hormones that control moulting, development from pupa to adult and other processes.

Instar A stage in the development of a larva between two moults.

Integrated control The manipulation of pest populations using any or all control methods in a sound ecological manner.

Interplanting The planting of one crop within another for the purpose of trapping pest insects.

Key pest An insect that is routinely present sometime during the growing season and that causes economic damage.

LC_{50} The lethal concentration of a substance, in air or liquid, for 50% of the test organisms.

LD_{50} The dose of a toxicant that produces 50% mortality in a population. LD_{50} values are used in determining mammalian toxicity (usually oral toxicity) and are expressed as milligrams of the toxicant per kilogram of body weight (mg/kg).

Larva (pl. larvae) The growing, worm-like stage of insects with a complete metamorphosis. Also, the newly hatched six-legged stage of mites and ticks.

Life cycle The complete succession of events in the life of an organism.

Maggot The growing stage, or larva, of a fly.

Mandibles The heavy pair of biting or chewing organs in an insect's mouth.

Mesothorax The middle section of the thorax of an insect that bears the top wings and the middle pair of legs.

Metamorphosis Any conspicuous changes in form or structure during the growth of animals.

Metathorax The last division of the thorax of insects, bearing the second pair of wings and the third pair of legs.

Microbial insecticide A microorganism that is applied in the same way as a conventional insecticide to control an existing pest population.

Mutagen Substance causing genes in an organism to mutate or change.

Nymph The growing stage of insects with an incomplete (gradual) metamorphosis.

Oral toxicity Toxicity of a compound when ingested. Usually expressed as number of milligrams of chemical per kilogram of body weight that kills 50% of the animals. The smaller the number, the greater the toxicity.

Organochlorine insecticide Chlorinated hydrocarbon, such as DDT, dieldrin, chlordane, BHC, lindane.

Organophosphate Class of insecticides derived from phosphoric acid esters.

Ovicide A chemical that destroys an organism's eggs.

Oviparous Reproducing by laying fertilised eggs.

Oviposition The act of laying or depositing eggs.

Ovipositor An organ, present in many insects, specialised for depositing eggs in plant structures or in the ground.

Ovoviviparous Producing living young from eggs that may hatch within the female's body.

Parasite Any organism that lives on or in the body of another living organism and obtains nourishment from it.

Parasitoid An insect that feeds in or on another organisms eventually destroying it.

Pathogen Any disease-producing organism or virus.

Parthenogenesis The production of young from unfertilised eggs.

Persistence The quality of an insecticide to persist as an effective residue, owing to its low volatility and chemical stability. Certain organochlorine insecticides are highly persistent.

Pesticide Any substance used for controlling, preventing, destroying, repelling or mitigating any pest. Includes fungicides, herbicides, insecticides, nematicides, rodenticides, desiccants, defoliants and plant growth regulators.

Pheromones Chemical substance which when released by an animal influences the behaviour of other individuals of the same species, e.g. sexual attractants in insects.

Phytotoxic Injurious to plants.

Point sample A method of sampling for insects designed to relate the number of insects or their damage to the number of plants and/or plant parts per unit area.

Polyphagous Feeding on a wide range of organisms.

Predator A predator is an animal that feeds on other animals. Insect predators usually require several types of prey to complete their development.

Preference Insect response to plant characters that leads to the use of a particular plant or variety for oviposition, food or shelter.

Prolegs The fleshy, unsegmented leg-like structures on the abdomens of some larvae.

Pronotum The dorsal sclerite of the prothorax.

Prothorax The first segment of the thorax, which bears the first pair of legs

Pupa (pl. pupae) The stage during which an insect with complete metamorphosis is transforming from the larval to the adult stage.

Puparium (pl. puparia) The darkened and hardened last larval skin of a maggot, which protects the pupa of true flies.

® Registered trade name.

Random sampling The most commonly used method of sampling for insects in which samples are taken at random to determine insect numbers or damage levels.

Re-entry intervals That time-interval, required by law, between the application of certain hazardous pesticides to crops and the exposure of workers to treated crops.

Residue Trace of a pesticide and its metabolites remaining on or in a crop, soil or water.

Resistance (insecticide) Natural or genetic ability of organisms to remain unaffected by a toxicant.

Sclerite A hardened body-wall plate bounded by sutures.
Sclerotised Hardened dense condition of cuticular structures.
Secondary pest A pest which usually does little, if any, damage but can become a serious pest under certain conditions, e.g. if insecticide applications destroy its natural enemies.
Segmented Divided into distinct parts.
Selective insecticide One which kills selected insects, but spares most other organisms, including beneficial species. This is achieved either through differential toxic action or through the way in which insecticide is used.
Sequential sampling A method of sampling insects that requires continued sampling until a pre-established upper or lower infestation level is found.
Sex lure A synthetic chemical which acts like the natural attractant (pheromone) for one sex of an insect species.
Spiracles The external openings to the breathing organs of insects and related forms.
Sternum (pl. sterna) The ventral body sclerite of an insect.
Strip cutting A term used to denote the practice of harvesting alternate borders of a crop, such as alfalfa, so that some partially-grown crop is maintained in the field at all times.
Stylets The slender, hollow, piercing and sucking organs of insects and nematodes that feed on plant sap.
Surfactant Ingredient of a pesticide formulation that alters the surface of the particles or droplets, so improving its properties (wetting agent, emulsifier, spreader).
Suspension Finely-divided, solid particles dispersed in a liquid.
Synergism Increased activity resulting from the effect of one chemical on another
Systemic Compound that is absorbed and translocated throughout the plant or animal.
Tarsus (pl. tarsi) The jointed 'foot' that bears the claws of an insect.
Tergum (pl. terga) The dorsal body sclerite of an insect.
Thorax That portion of an insect's body which lies between the head and abdomen and bears the legs and wings.
Tibia (pl. tibiae) The long, slim segement of an insect's leg to which the tarsus is attached.
Tolerance (residue) Amount of pesticide residue permitted by regulation to remain on or in a crop. Expressed as parts per million (ppm).
Tolerance (host-plant resistance) A basis of resistance in which the plant variety shows an ability to grow and reproduce or repair injury in spite of supporting a population of insects that would damage a susceptible variety
Toxicant A material that is poisonous.
Trade name (Trade mark, proprietary name, brand name) Name given to a product by its manufacturer or formulator, distinguishing it as being produced or sold exclusively by that company. Denoted in this book by ®.
Trap cropping A method of controlling pest insects by planting a more favoured host crop near the main crop and thereby preventing their movement into the main crop.
Vector An organism, such as an insect, that transmits pathogens to plants or animals.

Suggestions for further reading

1. Batra, S. W. T. 1982. *Biological control in agroecosystems*. Science **215**: 134–139
2. Barfield, C. S. & Stimac, J. L. 1980. *Pest management: an entomological perspective*. BioScience **30**: 683–689.
3. Busvine, J. R. 1975. *Arthropod vectors of disease*. London: Edward Arnold.
4. Debach, P. 1964. *Biological control of insect pests and weeds*. New York: Reinhold Publ. Crop.
5. Debach, P. 1974. *Biological control by natural enemies*. Cambridge: Cambridge University Press.
6. Flint, M. L. & Vanden Bosch, R. 1981. *Introduction to integrated pest management*. Net York: Plenum Press.
7. Hill, D. S. 1983. *Agricultural insect pests of the topics and their control*. 2nd ed. Cambridge: Cambridge University Press.
8. Huffaker, C. B. 1971. *Biological control*. New York: Plenum Press.
9. Huffaker, C. B. & Messenger, P. S. 1976. *Theory and Practice of biological control*. New York: Academic Press.
10. Kumar, R. 1984. *Insect pest control*. London: Edward Arnold.
11. Metcalf, C. L., Flint, W. P. & Metcalf R. L. 1962. *Destructive and useful insects*. 4th ed. New York: McGraw-Hill.
12. Metcalf, R. L. & Luckmann, W. H. 1982. *Introduction to pest management*. 2nd ed. New York: John Wiley & Sons.
13. Ordish, G. 1976. *The constant pest*. New York: Charles Scribers' Sons.
14. Pyenson, L. L. 1980. *Fundamentals of entomology and plant pathology* 2nd ed. Connecticut: The Avi publishing Com, Inc.
15. Roberts, D. A. 1978. *Fundamentals of plant-pest control*. W. H. Freeman and Company.
16. Samways, M. J. 1981. *Biological control of pests and weeds*. London: Edward Arnold.
17. Van den Bosch, R. & Messenger, P. S. 1973. *Biological control*. In text Press Inc.
18. Van Emden, H. F. 1974. *Pest control and its ecology* London: Edward Arnold.
19. Watson, T. F., Moore, L. & Ware, G. W. 1975. *Practical insect pest management*. W. H. Freeman and Co.
20. Woods, A. 1974. *Pest control: a survey*. McGraw-Hill Book Co. (U. K.) Ltd.
21. Youdeowei, A. & Service, M. W. 1983. *Pest and vector management in the tropics*. London: Longman Group Ltd.

Index

References to page numbers of illustrations are in italics.